RHYTHMIC AND SYNTHETIC
PROCESSES IN GROWTH

THE FIFTEENTH SYMPOSIUM OF
THE SOCIETY FOR THE STUDY OF
DEVELOPMENT AND GROWTH

Executive Committee, 1956
K. V. Thimann, Harvard University, President
M. V. Edds, Jr., Brown University, Secretary
R. W. Briggs, Indiana University, Treasurer
J. T. Bonner, Princeton University
Harriet B. Creighton, Wellesley College
J. D. Ebert, Carnegie Institution of Washington, Baltimore

Rhythmic and Synthetic Processes in Growth

THEODORE T. PUCK · R. DULBECCO

RICHARD M. KLEIN · DAVID M. PRESCOTT

COLIN S. PITTENDRIGH AND VICTOR G. BRUCE

E. BÜNNING · H. GAFFRON · HAROLD F. BLUM

BERNARD L. STREHLER · HARLOW SHAPLEY

EDITED BY DOROTHEA RUDNICK, 1907 —

WITHDRAWN
FAIRFIELD UNIVERSITY
LIBRARY

PRINCETON, NEW JERSEY
PRINCETON UNIVERSITY PRESS
1957

FAIRFIELD UNIV.
MAY 12 1958
LIBRARY

Copyright, © 1957, by Princeton University Press
London: Oxford University Press

All rights reserved

L. C. CARD 55-10678

Printed in the United States of America
by Princeton University Press at Princeton, New Jersey

FOREWORD

DIDEROT : *Le soleil eteint, qu'en arrivera-t-il? Les plantes periront, les animaux periront, et voila la terre solitaire et muette. Rallumez cet astre, et a l'instant vous retablissez la cause necessaire d'une infinité de generations nouvelles entre lesquelles je n'oserais assurer qu'a la suite des siecles nos plantes, nos animaux d'aujourd'huy se reproduiront ou ne se reproduiront pas.*

D'ALEMBERT : *Et pourquoi les memes elemens epars venant a se reunir, ne rendroient-ils pas les memes resultats?*

DIDEROT : *C'est que tout tient dans la nature, et que celui qui suppose un nouveau phenomene ou ramene un instant passé, recrée un nouveau monde.*

LE RÊVE DE D'ALEMBERT

THE Fifteenth Symposium of the Society for the Study of Growth and Development was held at Brown University, Providence, Rhode Island, on July 18–20, 1956. The Society is deeply indebted to the National Science Foundation which subsidized the Symposium and to Brown University and the local committee who made the occasion so agreeable and so fruitful. Most sincere thanks are due all those, within and without the Society, who contributed to this success, as well as to all those, both within and without the Princeton Press, who have brought the present volume to completion.

The program was arranged in the form of three subsymposia, each coherent within itself. The subject of the first day was the recent great advances in tissue culture methods which have opened such vast new experimental possibilities in cell biology. Theodore T. Puck discussed his innovations in the field of clonal cultures of animal cells; R. Dulbecco his progress into the stimulating problems of virus reproduction; and Richard M. Klein the equally active field of plant tissue cultures.

The second day was occupied with problems of cyclic activity: the growth-division cycle in Amoeba (David M. Prescott); time-measurement by organisms (Colin S. Pittendrigh); and diurnal rhythms in vascular plants (E. Bünning). The last group of speakers—H. Gaffron, Harold F. Blum, B. L. Strehler, and Harlow Shapley—devoted themselves to questions of biochemical evolution, and indeed, as the reader will observe, did not hesitate to take off in the grand manner to primordial time and galactic space. The morning and the evening were the third day.

35406

The Executive Committee feels that the Fifteenth Symposium marks an appropriate point for a backward look at the Society's efforts and progress thus far. This volume includes, therefore, an index, by authors, of articles published in the first fifteen Growth Symposia, beginning with the first one in the summer of 1939.

DOROTHEA RUDNICK

Yale University

CONTENTS

CONTENTS

RHYTHMIC AND SYNTHETIC
PROCESSES IN GROWTH

I. THE MAMMALIAN CELL AS MICROORGANISM

BY THEODORE T. PUCK[1]

I. INTRODUCTION

MAMMALIAN biology for the most part has been the biology of the *macro*organism—study of the body as an architectural structure with specialized functions lodged in various tissues and organs. The cell has been recognized as the ultimate unit of this structure, and the study of individual cell morphologies has contributed enormously to the understanding of how different tissues are constituted so as to perform their specific functions. Yet in no case is it clear to what extent a given cell is autonomous and to what extent it is governed by its status as a sub-unit of a particular tissue, which is itself tightly integrated into the over-all economy of the body.

Detailed morphological studies of individual cell types have yielded exceedingly important data, much of which has become indispensable in many procedures of medicine. A wealth of biochemical information has also accumulated describing the function and distribution of enzymes, nucleic acids, specific proteins and other molecules in specific regions of special cells. But there remain largely unanswered questions like the following, relating the metabolism of the individual cell to the total body economy:

1. What proportion of the cells of various normal tissues are capable of independent growth and multiplication in isolation if furnished with the proper nutrient environment? I.e., which cells are capable of existence as true microorganisms, and which are dependent for continued metabolism and reproduction on close association in a community with other cells of similar or different constitution?

2. Which of the different cell types arising during normal embryonic differentiation represent changes in genetic constitution which are relatively fixed, and to what extent are these merely adaptive responses which all cells may exhibit when properly stimulated but which affect different cells differently, simply because of factors depending on physico-chemical, temporal and spatial relationships?

[1] Department of Biophysics, Florence R. Sabin Laboratories, University of Colorado Medical Center, Denver. Contribution No. 47.

3. May mammalian cells undergo genetic modification by the processes of transformation, transduction, or direct sexual exchange of nuclear or cytoplasmic genetic determinants?

4. What is the distribution of nuclear genes in the mammalian chromosomes; what is the frequency of processes like crossing-over and spontaneous gene mutation; and how do mutagenic agents like high energy radiation affect the growth potential and genetic constitution of individual mammalian somatic cells?

5. What is the chromosomal and genic basis of the various defects responsible for mammalian genetic diseases?

6. To what extent do the processes like cancer formation and ageing represent changes in the genetic or the physiologic components of the cell?

Answers to questions like these require means for study of the growth and genetic processes of mammalian cells in a manner similar to that which has been so rewarding in microorganisms like *Neurospora* and *E. coli*. An experimental program directed toward this end is in progress in our laboratory.[2] Study was initiated to obtain quantitative growth of single cells, i.e. to find conditions such that practically every cell of a mammalian population will reproduce in isolation to form colonies of any desired size. It must be possible to carry out such quantitative operations rapidly and simply, and on a very large scale, to permit discovery and isolation of the genetic mutants which can serve as markers to illuminate hereditary processes. Since gene mutations generally occur with frequencies in the neighborhood of one in a million, methods requiring separate manipulation of each individual cell cannot be employed. Hence, attention was devoted to devising methods for growing single mammalian cells into colonies in a fashion completely analogous to the growth of single bacteria plated on solid nutrient media.

Growth of mammalian cells in vitro by conventional tissue culture methods is by this time a commonplace operation. These procedures, requiring as they do an inoculum of approximately 10^5 cells in order for self-sustaining growth to be initiated (Earle et al., 1951), make it impossible to determine whether, under the conditions employed, a cell *community* is needed to initiate continuing growth, or whether only a rare cell is capable of reproduction. In the latter case, the magnitude of the required inoculum would be a measure of the number of cells needed

---

[2] The experiments described are the results of a joint program by the author and the following co-workers: Steven Cieciura, Harold Fisher, and Philip Marcus, who are candidates for the Ph.D. degree.

to insure the presence of at least one reproducing individual. Earle and his co-workers who emphasized the need for cells to "condition" the conventional tissue culture media in order to make them able to support growth, succeeded in inducing self-sustaining multiplication from several different mouse cells, by introducing single cells into fine capillaries which were then sealed and incubated (Sanford, Earle, and Likely, 1948). These critical experiments at once established that at least some single mammalian cells can multiply in isolation, and permitted the establishment for the first time of true clonal stocks. (A clone is a population all of whose members have descended from the same single member by asexual reproduction.) However, only a small proportion of the cells so isolated did multiply. Further, the necessity for separate manipulation of each individual cell precludes the routine screening of large numbers of individuals which is prerequisite for systematic genetic studies. Growth of single *ascites* tumor cells into large populations has been demonstrated in vivo (Yoshida, 1952) but this technique also is not readily applicable to the large populations required by genetic studies. Moreover, since this growth takes place in vivo, the possibility of genetic interaction between the inoculated cell and the host always adds uncertainties to the interpretation of such experiments.

II. QUANTITATIVE GROWTH OF SINGLE CELLS

The successful plating of single mammalian cells on petri dishes under conditions leading to colony formation with an efficiency close to 100% was achieved first with cells of the HeLa Strain (Scherer, Syverton, and Gey, 1953), originally isolated from a human cervical carcinoma. The technique for this single cell plating has been published in detail (Puck, Marcus, and Cieciura, 1956). The only significant difference between this method for plating animal cells and that which is the basis of quantitative bacteriology, lies in the additional necessity for dispersing into single cells the aggregates in which mammalian cells characteristically tend to grow. If the reproductive potential of each single cell is to be preserved, this procedure must be carried out without trauma. Either trypsin or a chelating agent may be used for this purpose, but it is imperative to follow faithfully a procedure which insures complete dispersal without significant damage to the cells (Puck, Marcus, and Cieciura, 1956). If the medium available is nutritionally adequate, it is sufficient simply to pipette an inoculum containing the desired number of dispersed cells onto a petri dish containing several cc of the growth medium. The cells attach to the glass, and since these are initially well mixed in the medium, and usually present in numbers less than 200, each cell is sufficiently far from its neighbors to produce a discrete colony of about 2 mm in diameter after 9 or 10 days incubation. On Plate I, 1-3, are presented some typical colonies arising from the plating of single cells originating from a variety of human tissues.

While this method is satisfactory for some cells in the Complete Growth Medium which we have adopted as a standard in our laboratory (Marcus, Cieciura, and Puck, 1956), other cells fail to multiply when plated therein after dispersal. Yet these same cells can multiply indefinitely in the same medium if seeded in a large inoculum instead of in the single state. For example, a massive inoculum of HeLa S_3 cells will multiply rapidly and continuously in a basal medium to which no cholesterol has been added, but single cells require this metabolite added to the medium, in order to reproduce. This constitutes a model situation in which cell reproduction requires cooperation between several individuals either through the need to modify the medium as Earle has suggested, or possibly also through some other more direct type of cell-cell interaction.

The alternative plating technique devised to permit colony formation from single cells which fail to multiply in the usual nutrient media utilizes a "feeder" system of cells whose own reproduction has been terminated by a previous exposure to x-irradiation (Puck, Marcus, and Cieciura, 1956). The use of high energy radiation is particularly well suited to this purpose because of its enormously greater potency in blocking reproductive as opposed to other metabolic activities. We have demonstrated that the mean lethal dose of x-rays for a mammalian cell involves an energy absorption equivalent to a temperature rise of only 0.0001°C (Puck and Marcus, 1956). It was demonstrated that the addition of such x-rayed cells to a plate seeded with a measured inoculum of single cells which by themselves show no growth, results in 100% efficiency of colony formation. Hence, the use of such a "feeder" system makes possible extension of this plating technique even to cells otherwise deficient in the ability to grow as independent microorganisms in the media available (Plate I, 4).

Experiments have demonstrated that cells differ greatly in their capacity to act as feeders. Thus, while the irradiated HeLa cell can act as an effective feeder for normal HeLa cells in a nutritionally deficient medium, it can not feed human fibroblasts. These latter, however, can feed either themselves or the HeLa cell. Thus, the use of a feeder system furnishes a new investigative tool for various phases of cell–cell interaction and, indeed, suggests applications that may fruitfully be applied to studies with molds and bacteria.[3]

The advantages of a quantitative method for producing growth of single cells by a rapid and convenient plating procedure may be summarized: (1) It provides means for accurate titration of the reproductive powers of the individual cells of a population. Thus it furnishes a much more potent tool for study of the dynamics of cell growth and the effects thereon of physical and chemical agents, including nutritional and toxic factors. (2) It makes possible ready recognition and isolation of genetic mutations, and thus furnishes the means for quantitative study of an enormous variety of genetic processes in animal cell populations. (3) With the aid of the feeder system it permits new kinds of experimentation on cooperation and antagonisms in the interaction of cells with other cells of the same and different kind.

The remainder of this discussion will deal with representative experiments which have been carried out in our laboratory, illustrating applications of this plating technique to specific problems.

A. Establishment of clones from different tissues. A series of experiments is in progress testing how wide a variety of human cell types can be plated by these means. Cells of both epithelial and fibroblastic morphology have already been plated in this way with efficiencies in the neighborhood of 50–100%. The tissues represented include human skin, liver, conjunctiva, appendix, kidney, spleen and bone marrow. Good results have been obtained with cells recently isolated from human subjects as well as those previously cultivated for long periods in tissue culture. So far, we have not encountered a human cell type which has absolutely failed to grow under conditions of plating as single cells— i.e. which has not retained the capacity to be an independent microorganism. It becomes of primary importance to determine which, if any, of the body cells have indeed irreversibly lost this potentiality. Current studies include measurement by this same means of the growth potentials of cells from a variety of human tissues taken from subjects of different ages, and exhibiting a variety of different disease conditions.

B. Study of agents which affect growth—the action of x-rays on mammalian cells. This technique has been applied to analysis of the action of various agents on the cellular growth process. A typical

---

[3] It is of interest that the basic principle of feeder cells had been used for the cultivation of single plant cells (Muir, Hildebrandt, and Riker, 1954). These investigators did not use irradiated cells for their feeder system but rather laid a filter paper inoculated with single plant cells on top of a macroscopic segment of living plant tissue, the molecular exchange then progressing by diffusion through the filter paper.

growth curve obtained from a single plate by counting the cells per colony in 20 or more colonies after various incubation periods is pre-

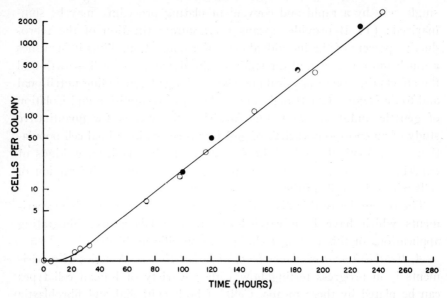

Fig. 1. Typical growth curve of single HeLa S3 cells when plated by the technique here discussed. In optimal medium, an almost identical curve is obtained from cells of conjunctiva, liver and kidney.

sented in Fig. 1. Such curves have been shown to be highly reproducible for a given cell in a given medium, even after the lapse of more than a year (Puck, Marcus, and Cieciura, 1956). It is evident, then, that a curve derived from such a single plating procedure can yield three types of quantitative information about the nature of the growth process: (1) the percent of the individuals of the population capable of forming colonies under the particular conditions employed; (2) the average lag period before growth begins, which presumably represents the time needed for adaptation to the given environment; and (3) the generation time—i.e. the time taken for the population to double when the colony is in its logarithmic growth phase. Thus, by providing different quantitative measures of various aspects of cell growth, this technique affords more comprehensive analysis of the action of agents which can affect the growth process. For example, it has been demonstrated that decreasing the concentration of human or porcine serum from 15% to 3% in the Growth Medium, reduces the growth rate of S3 cells but does not change the plating efficiency—i.e. the percentage

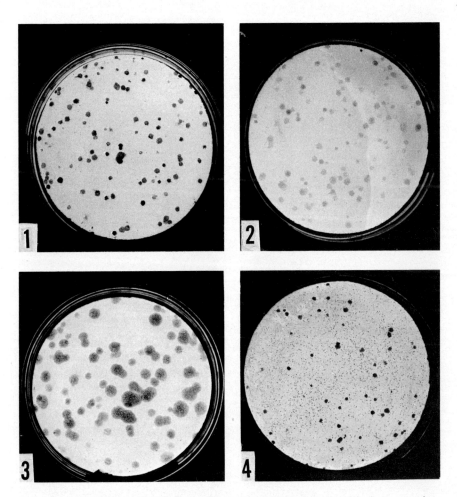

PLATE I. 1–3. Typical colonies arising from the plating of single cells from a variety of human tissues. Each plate was inoculated with 100 cells of the type shown. All photographs are actual size. 1. Carcinoma of the cervix (HeLa). 2. Normal human liver. 3. Normal human conjunctiva. 4. Colonies developed from single cells of a deficient HeLa mutant which does not grow in any medium available unless a layer of x-irradiated feeder cells is present. The gray speckled background is the layer of nonreproducing feeder cells.

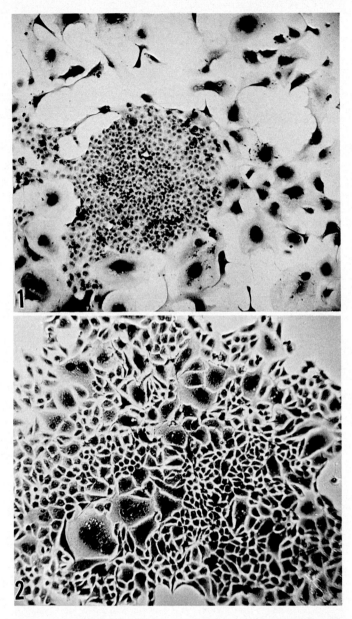

PLATE II. 1. Photomicrograph of a normal colony of HeLa cells which is growing up among irradiated HeLa cells which are growing but not multiplying, and so are becoming giants. The giants can achieve spread diameters that are about 10 times that of a normal cell. (Puck and Marcus, 1956.) ×25. 2. Typical colony produced from single cell of the bizarre mutant of the HeLa S3. This morphological mutant is characterized by the production of colonies containing a large proportion of monster cells. This mutant was produced by x-irradiation of the clonal strain of S3 cells. ×54.

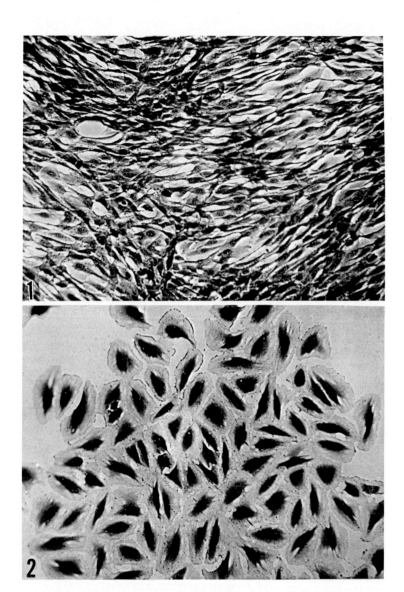

PLATE III. Demonstration of genetic differences in morphology of human cells from two different sources when single cells are grown into colonies in a medium of constant composition, which contained 20% human serum, 5% embryo extract, and 75% Hanks saline. 1. Fibroblast-like cells isolated from normal human foreskin. ×115. 2. Epithelium-like cells isolated from normal human liver. ×115.

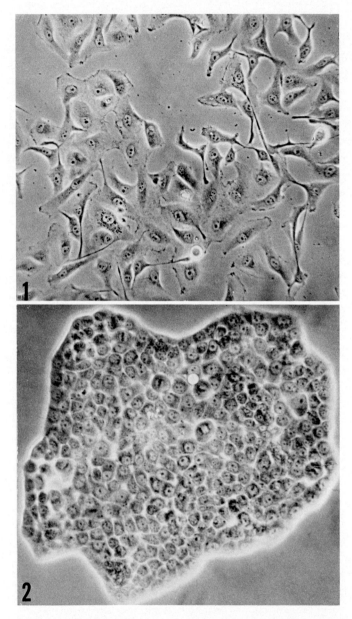

PLATE IV. Demonstration of a nongenetic morphological change when human cells from the same clone are grown in two different media. (Puck, Marcus, and Cieciura, 1956.) This change is reversible. 1. Fibroblast-like colony developing from a single S3 cell growing in the presence of 20% human serum. 2. Epithelium-like colony developing from an identical cell, but grown in horse or porcine but not human serum.

of the single cells which can form colonies. If the serum concentration falls below 2%, then the number of individuals which can reproduce also drops sharply. Similar studies are in progress on the effects of temperature, and a variety of drugs, and other agents of chemical and biological origin.

It is of interest that every cell type with which we have used this technique, originating from a variety of human tissues, and including both normal and carcinomatous specimens, has exhibited the same maximum growth rate, equivalent to a generation time of 18-20 hours (Marcus, Cieciura, Puck, 1956). All these cells also exhibit a similar mitotic time of about 40–50 minutes under these conditions. These observations suggest a limiting maximal reproductive rate for human cells, governed primarily by some rate-limiting process during the inter-kinesis period.

One study which merits special consideration involves the action of high energy radiation on the growth of mammalian cells (Puck and Marcus, 1956). Other techniques used in attempt to quantitate the destructive action of ionizing radiation on mammalian cells have yielded values ranging from several hundred to several hundred thousand roentgens as the mean lethal dose, a divergence so great as to render the data almost meaningless for many purposes where accurate information is needed. The present plating technique makes possible exactly the same kind of precise, survival-curve determination for the reproductive potential of mammalian cells as has become standard for bacteria. Plates inoculated with known numbers of single cells were irradiated with a series of different doses of x-irradiation, then incubated in the standard manner. The survivors which had retained the ability to form macroscopic colonies were then counted. A typical survival curve for the clonal HeLa cell is presented in Fig. 2. The shape of this curve indicates a 2-hit process as the one responsible for destruction of the reproductive capacity, and combined with other data, leads to the con-clusion that the critical events have taken place in the cellular genetic apparatus (Puck and Marcus, 1956). The value for the mean lethal dose obtained from this curve is 96r, an astonishingly low one. The original study was carried out on a cell from the HeLa carcinoma of the cervix, and it was at first thought that this cell might have a radiation sensi-tivity far greater than that of cells from normal tissues. However, repeated determination of the survival curve on cells originating from normal tissues, has yielded values for the lethal dose which in no case so far exceed that of the HeLa cell by more than 20-30%.

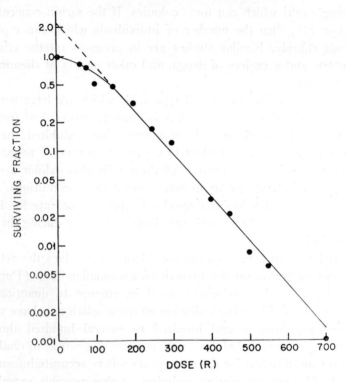

Fig. 2. Survival of reproductive capacity of the HeLa cell as a function of x-ray dose. (Puck and Marcus, 1956.)

These data throw some light on radiation effects in man and other mammals. It has been difficult to understand why the mean lethal dose of total body radiation for man should be so little as 400r, particularly since so many cell and tissue functions like muscle contraction, nerve conduction, and oxygen consumption are almost unaffected by exposure to many thousands of roentgens. The demonstration that damage to the genetic apparatus of various human cells can occur with a mean lethal dose of 96r illuminates at least one mechanism which undoubtedly contributes to the pathology of the acute radiation syndrome. Damage to the cell genes will, in general, not become manifest until the time for cell multiplication arrives, at which point cell reproductive failure will ensue, or abnormal progeny will be produced. Therefore, those tissues like the skin, the lining of the gastrointestinal tract, and the blood-forming organs, containing cells which are in constant multiplication, should exhibit considerable radiation damage at relatively low doses, a principle which has long been recognized by radiologists. The genetic nature

[10]

of such cell damage could also furnish an explanation for the lag which separates the radiation experience and the development of pathological symptoms, which is another of the most characteristic features of radiation injury. Damage to cellular genetic elements would not become evident until the particular cell involved reaches its time for mitosis, and, indeed, we have found that often such injury does not find expression until the original irradiated cell has undergone 4 or 5 successive mitotic divisions. At the end of this period, all of the progeny lose the ability to multiply further. This sequence of events which could only be delineated by a methodology which permits accurate tracing of the history of each irradiated cell and its progeny, affords a very persuasive picture of the nature of the lag in appearance of radiation pathology in the whole mammalian organism. Undoubtedly other factors contribute to the extraordinarily complex clinical course of events included by the radiation syndrome. There can be no doubt that direct physiologic damage to certain cells also plays a role. One apparently non-genetic injury sustained by irradiated cells in the region of 200-400r is an inhibition in growth which appears to be self-reparable (Puck and Marcus, 1956). Studies analyzing these effects and their dependence on various metabolic parameters are continuing. It is of interest, in this connection, that the toxic cellular effect we have studied cannot be due to production of specific molecular inhibitors in the serum-containing medium. This was demonstrated by irradiating the medium itself with many thousands of r, equivalent to more than 10 lethal doses. Such medium, when added to normal cells, supported their growth in exactly the same manner as that of medium never exposed to x-rays.

The fate of the cells whose ability to reproduce indefinitely has been destroyed by radiation is of interest to the present discussion. As was indicated in a previous paragraph, the dose necessary to accomplish this destruction represents so tiny an amount of energy as to make it certain that an exceedingly small fraction of the atomic bonds present in the molecules constituting the cell, have been broken. So small an amount of chemical change, while sufficing to block reproduction, must leave intact the great majority of the cell machinery. Hence, one would expect such cells to continue metabolizing and carrying out many of the functions of completely normal cells. Such is indeed the case. These cells continue to assimilate nutrients from the medium, but being unable to reproduce, continue to increase in size, achieving huge proportions and becoming readily visible to the naked eye (Plate II, 1). These giant cells may attain stretched diameters on glass of almost 1 mm, and constitute fascinating metabolic structures. Like normal cells, they attach to glass, can be redispersed by trypsin, and will reattach to glass in the proper medium. They metabolize glucose, lowering the pH of standard nutrient medium at approximately the same rate as a similar mass of normal cells. They are susceptible to attack and destruction by mammalian viruses—indeed, more sensitive in this respect than normal cells, possibly because of increased surface area (Puck and Marcus, 1956), and readily take up vital stain. The ability of such cells to constitute a "feeder" layer to promote the growth of single, normal cells has already been implied in the previous discussion, since the feeder layers produced are composed of these giant forms. Methods have been described whereby a cell population of any desired number of pure giants can be prepared for experimental study. Detailed investigations of these forms would appear to be rewarding at the fundamental level, and also to have applications to problems of clinical radiology, as for example, in situations like those studied by Graham (Graham and Goldie, 1955), where giant cell appearance has been used as an empirical index of the ability of carcinoma of the cervix to be treated by radiation therapy.

C. The genetics of mammalian somatic cells.

(1) *Phenotype and genotype.* The cell plating method here discussed furnishes a particularly convenient method for isolating clones of any cell which can be so cultivated. The procedure which has been described requires only a few minutes to pick for subsequent cultivation, any of the discrete colonies developing on the routinely produced plates. Since each colony has arisen from a single cell, it automatically constitutes a clone. As in the corresponding bacteriological procedure, we have adopted the practice of passing each clone through at least two successive single cell isolations, in order to insure against contamination with cells from neighboring colonies.

The first question which one might wish to ask with the aid of these procedures concerns the possible existence of genetically different cells in the same mammalian individual—a problem of fundamental importance in orienting investigational attitudes towards processes like embryonic differentiation and cancer formation. The concept of a stable mutation has developed an independent operational meaning in microbiological procedures which has demonstrated its usefulness. A stable mutant has come to be accepted as a cell descended by asexual reproduction from a parent cell, but with demonstrably different behavior in a controllable situation. This difference must be demonstrated by all of the new subclones arising from multiplication of isolated single cells of the mutant. Also implicit in the operational definition which has gradually developed out of the fruitful work of the last two decades in the genetics of microorganisms is the concept of stability: a true mutation in this sense does not occur spontaneously in more than a tiny fraction of the population, i.e. less than 0.1%. Similarly, a mutant population must not spontaneously revert to the parental type to an extent greater than about 0.1% in any generation. These frequencies may be drastically changed artificially by the action of an agent specifically demonstrated to affect genetic constitution in a variety of cell types, but it is imperative that the property under study remain stable in successive lines of progeny long after withdrawal of the inducing agent. While this definition leaves a good deal to be desired, and hopefully will shortly be replaced by a more chemically specific one, it is useful as a laboratory guide. Thus, for example, it excludes an adaptation process like enzyme induction, which gradually disappears from the progeny of an induced population after the inducing agent is withdrawn. It includes, however, mutations produced by transforming principles, as well as the fascinating series of alternative genetic changes which can be triggered in *Paramecium*, as described by Sonneborn and his colleagues. It is important to note that once a stable mutation has been demonstrated, there still remains to be determined whether it is a specific gene mutation, a chromosomal alteration of one or another kind, a change in cytoplasmic genetic determinants, and whether transforming principles, transducing viruses, or other as yet unrecognized factors are involved as causal or contributing factors.

By this definition which we have adopted, any two demonstrably different cell strains originating from the same mammalian individual, but which retain their differences when continuously cultivated from single cells, are to be regarded as mutants. These cells originated mitotically from the same fertilized egg, so were originally members of the same clone. It is crucial, however, to insist that mutant demonstration cannot be accepted without proof of the persistence of the difference in behavior of the two cell strains after single cell isolation. For example, the term "mutation," as defined here operationally, has meaning only when applied to an independent microorganism. The mammalian cell, demonstrated to be a microorganism, must then conform to the same operational tests if the concepts of microbiological genetics are to be used meaningfully. Thus, the demonstration that two cell types isolated by macroscopic biopsy from the same individual, and maintained indefinitely in tissue culture by passage of macroscopic inocula, does not suffice to demonstrate them as mutants. The original biopsy material may contain cells of the same genotype but widely differing in their phenotype because of differences in the adaptive states and the spectrum of induced enzymes and other metabolic agents present in each cell. Cultivation of such an aggregate of cells by the techniques of conventional tissue culture would offer opportunity for such cells to interact on each other, and perpetuate these differences.

One of the most obvious tests for the existence of true genetic differences in mammalian cells is that contained in the morphologic distinction between cells of fibroblastic and epithelial origin. While the difference between these two cell types has always been accepted by embryologists as an extremely deep-seated one, no test has been put forth demonstrating whether genetic differences in the specific sense here employed, are involved. When plated as single cells under identical conditions, in the same medium, both cell types produce morphologically distinct colonies, readily recognizable as fibroblastic or epithelial type, respectively. Typical photographs are illustrated in Plate III. It may be concluded that the difference in morphology of these two cell types is a true, stable, genetic trait. Hence, the conclusion appears inescapable that the process of mammalian differentiation involves true changes in the genetic constitution of the daughter cells in a manner which, at present at least, is operationally indistinguishable from that familiar from studies of forms like *E. coli, Pneumococcus,* and *Paramecium.*

Not every morphologic variation from spindle-shape to cuboidal cells is evidence of a genetic change, however. Thus it has been found possible to alter the characteristic morphology of all the cells of a given clonal strain, so as to render them more epithelium-like or more spindle-shaped, simply by changing the constitution of the growth medium. Representative colonies of the same clonal stock grown in the two media are shown on Plate IV. The change is very marked, and is brought about by a constituent present in greater concentration in human serum than in bovine, equine, or porcine serum. In the presence of 20% of human serum, the cells grow as tightly stretched, spindle-shaped forms which are highly migratory, as evidenced by the relatively greater amount of space between the cells of such colonies as contrasted with the tightly packed array characteristic of the same cells grown in porcine serum. The change in either direction is completely reversible. Thus, cells which have grown in porcine serum to form the tight, cuboidal colony, will change their morphology to assume the highly stretched, migratory form, if the medium is changed after several days of incubation to one containing 20% or more of human serum. Similarly, the reverse morphological change can also be effected by switching from porcine to human serum after several days incubation.

Clonal stocks of epithelium-like cells, obtained from normal human tissues, also exhibit this change to the more highly spread condition when grown in human serum, but the magnitude of their response is much less striking than that of the carcinomatous HeLa. This difference in behavior may be related to the fact that the cells from normal tissues are normally more stretched, occupying a greater surface area, than the HeLa, in any medium (Marcus, Cieciura, Puck, 1956). The possible significance of this observation for problems of the cellular nature of malignancy remains to be explored.

Studies have also been carried on to determine the presence of spontaneously occurring mutants in human cells (Puck and Fisher, 1956). The first of such investigations was carried out with the parental HeLa cell population, originally isolated as a macroscopic biopsy and maintained for years by the procedures of conventional tissue culture, which involve transfer of macroscopic inocula. This cell strain has been considered to be particularly pure, because of its morphological uniformity when grown in vitro by standard techniques. However, when this parental strain was grown by the single cell plating method, the colonies which developed exhibited great heterogeneity in size and appearance, suggesting the existence of mutants within the population. This suggestion was strengthened when it was found that clonal stocks obtained by picking such individual colonies yielded, on single cell plating, colonies with highly regular and uniform morphology. Examination of the growth requirements of two such clonal stocks revealed distinct, heritable differences in their nutritional requirements for colony formation which make possible sharp differentiation between the two strains: Cells of the S3 clone grow with 100% efficiency in the standard synthetic nutrient medium plus 20% of the macromolecular fraction of human serum. The S1 clone shows no growth whatever under these conditions. The genetic stability of the difference in growth requirements of these two forms has been demonstrated by the fact that their behavior has remained unchanged throughout more than 100 generations of cultivation in our laboratory, which has included, in each case, two or more successive single cell isolation procedures. It is of further importance that the two cell types are indistinguishable morphologically, despite the completeness of their differentiation by growth in the selective media described. This demonstration that stable genetically different lines can exist in cells taken from the same tissue, and which are morphologically indistinguishable, further emphasizes the importance of study of tissue cell population dynamics by techniques capable of revealing such behavior.

(2) *Action of mutagenic agents.* With the establishment of stable genetic markers for somatic mammalian cells, it became possible to study both qualitatively and quantitatively the action thereon of mutagenic agents, a procedure for which urgent need has been raised by the unprecedented increase in human exposure to ionizing radiations which our generation is experiencing. Radiation-produced mutants of human cells have been produced which exhibit new nutritional requirements not displayed by the original parental stock. Studies are now current measuring radiation-induced frequency of mutation with respect to such nutritional requirements of a variety of human cells from normal and malignant tissues. While the biochemically deficient mutants are the ones receiving greatest emphasis in these studies, other types of radiation-induced genetic changes have also been observed. Figure 2 on Plate II illustrates one of the morphologic mutants of the S3 clone of HeLa produced by exposure to a dose of 800r. This mutant, which we have named *"bizarre,"* produces colonies from its single cells, containing an extremely high proportion of cell monsters. The clone was isolated originally from the survivors of the radiation, passed through two further single cell isolations, and is now cultivated routinely in bottles as a standard laboratory clonal stock.

These experiments demonstrate that the individual mammalian cell has a biology fully as complex and interesting as that of any other microorganism. Study of those properties which different human cells possess when grown in isolation must also illuminate the functions which require physical and chemical cooperation between cells for their fulfillment. The future promises great advances in the understanding of the dynamics of mammalian systems at both these levels.

BIBLIOGRAPHY

Earle, W. R., K. K. Sanford, V. J. Evans, H. K. Waltz and J. E. Shannon, Jr. 1951. The preparation and handling of replicate tissue cultures for quantitative studies. *J. Natl. Cancer Inst. 11*, 907–927.

Graham, R. M., and K. R. Goldie. 1955. Prognosis in irradiated cancer of the cervix by measurement of cell size in the vaginal smear. *Cancer 8*, 71–77.

Marcus, P. I., S. J. Cieciura and T. T. Puck. 1956. Clonal growth *in vitro* of epithelial cells from normal human tissues. *J. Exp. Med.*, 1956, in press.

Muir, W. H., A. C. Hildebrandt, and A. J. Riker. 1954. Plant tissue cultures produced from single isolated cells. *Science 119*, 877–878.

Puck, T. T. and P. I. Marcus. 1956. Action of x-rays on single mammalian cells. *J. Exp. Med. 103*, 653–666.

Puck, T. T., P. I. Marcus, and S. J. Cieciura. 1956. Clonal growth of mammalian cells *in vitro*. II. *J. Exp. Med. 103*, 273–284.

Sanford, K. K., W. R. Earle, and G. D. Likely. 1948. The growth *in vitro* of single, isolated tissue cells. *J. Nat. Cancer Inst. 9*, 229–246.

Scherer, W. F., J. T. Syverton, and G. O. Gey. 1953. Studies on the propagation *in vitro* of poliomyelitis virus. IV. Viral multiplication in a stable strain of human malignant epithelial cells (Strain HeLa) derived from an epidermoid carcinoma of the cervix. *J. Exp. Med. 97*, 695–709.

Yoshida, T. 1952. Studies on an ascites (reticulo endothelial cell?) sarcoma of the rat. *J. Nat. Cancer Inst. 12*, 947–969.

II. THE MULTIPLICATION OF RNA-CONTAINING ANIMAL VIRUSES

BY R. DULBECCO[1]

THE STUDY of viruses is of great interest to biologists because of the particular nature of viruses. A virus particle is in fact a relatively small mass of organic material which is endowed with two essential characteristics of living things: it can multiply and it has genetic continuity. Viruses are therefore a very favorable material for the study of the chemical and physical bases of growth and of heredity.

In the last two decades most of the fundamental work on the biology of viruses was confined to the bacterial viruses. Many important advances were made, due to the high accuracy that bacteriophage work allows and to the simplicity of the techniques involved. The necessity for fundamental work in other fields of virology was not strongly felt until recently, when important reasons appeared for considering other types of viruses with equal attention. The most compelling reason was the recognition that in many viruses, both of animals and of plants, ribose nucleic acid (RNA) appears to be the only nucleic acid, and most likely the carrier of the genetic information; it is known that, on the contrary, desoxyribose nucleic acid (DNA) is the genetic material of bacteriophages.

This difference between the two groups of viruses posed a new fundamental problem for biology. In fact, as long as RNA was found only in the cells, it could be visualized as having a secondary role, subordinate to that of the nuclear DNA. In the multiplication of an RNA virus, however, the RNA must play a primary role. Thus questions of the following type arise: Is there a dual mechanism of self-replication? Or, in spite of the existence of the two types of viruses, is the essential mechanism of self-replication unique? And, how different are the functional (genetic) properties of the two nucleic acids?

To attack these new problems a model system for the study of the multiplication of RNA viruses had to be chosen. In our laboratory some RNA-containing animal viruses were selected. This choice was determined by the fact that these viruses can be studied quantitatively in a satisfactory way both chemically and biologically.

A summary of the results of the work concentrated on the analysis

[1] California Institute of Technology, Pasadena.

of such systems during a four-year period will be presented in this paper. The work to be discussed was initially directed toward the solution of technical problems; an exact method of assay for animal viruses—the plaque method—was worked out. Then the work was concentrated on the investigation of the physiological, genetical, and chemical properties of a few selected systems.

A brief recapitulation of the technical developments will be useful. The plaque technique (Dulbecco, 1952) which is used to enumerate the infectious particles in a virus preparation is similar in principle to that used in bacterial virus work. A cellular layer one cell thick, called a monolayer, is first grown on the glass bottom of a petri dish; a small volume of the virus sample to be assayed, properly diluted, is deposited on the monolayer, which is later overlaid with a layer of nutrient agar. After a period of incubation at 37°C, round areas of necrosis, called plaques, develop on the monolayer. The plaques can be made more evident by staining the surviving cells with neutral red or the dead cells with trypan blue (Lwoff, 1956). In a variation of this technique a suspension of cells is used: the cells are suspended in molten agar mixed with the virus and poured on top of a layer of nutrient agar (Cooper, 1955).

It can be proved that, in general, a single virus particle in contact with the cellular monolayer is sufficient to give rise to a plaque. In brief, the evidence is the following (Dulbecco and Vogt, 1955): (1) there is a linear relation between the number of plaques and the amount of virus placed on the cell layer; this shows that not more than one viral unit indivisible on dilution is required to initiate a plaque; (2) the inactivation curves of the plaque-forming ability of several viruses under the action of several inactivating agents (ultraviolet light, x-rays, heat, antibody) is of 1-hit type; this shows that a plaque-forming unit contains generally not more than one active virus particle.

The efficiency of plating, i.e. the ratio of plaques produced by a virus sample to the number of particles countable with the electron microscope in the same sample, varies greatly in different systems. For the Newcastle disease virus this ratio may be of the order of 2 (Levine et al., 1953). For poliomyelitus virus, on the contrary, the particles to plaques ratio is of the order of 30 to 100 or higher (Schwerdt and Schaffer, 1955; Fogh and Schwerdt, 1956). The reason for discrepancy between number of particles and number of plaques is still unknown. Two hypotheses can be visualized: either all the particles are alike and have a rather small probability of initiating the infection of the cellular monolayer, or the particles that produce plaques are different from those that do not. Neither hypothesis can be ruled out at present. However, several observations can be summoned in favor of the second hypothesis (Dulbecco and Vogt, 1955); to these a new observation should be added, i.e. that the particles to plaques ratio varies only by a small factor (about 2) in a number of different cellular systems having nearly optimal efficiency of plating, such as *Rhesus* or *Cynomolgus* or *Erythrocebus* (Melnick, 1956) monkey kidney cells, human amniotic cells (Fogh and Lund, 1955), cells of various human cancers, such as the HeLa cells (Scherer et al., 1953) and the K.B. cells (Eagle, 1955).

Another relevant technical point is the extensive use of cell suspensions in the experimental work that will be discussed. The use of cell suspensions, on one hand, allows the determination of essential experimental parameters of virus infection (Dulbecco and Vogt, 1954), and, on the other hand, permits the isolation of the progeny of a single infected cell (Lwoff et al., 1955). The cells used in our work do not go spontaneously into suspension; trypsin or chelating agents are used for this purpose. These procedures do not affect the essential features of virus reproduction within the cells, but change some of its aspects. The yield of virus per infected cell decreases in the suspended cells to 20–50% and the shape of the growth curve is changed somewhat. Howes and Melnick (1956) have shown that monkey kidney cells infected by poliomyelitis virus release the virus faster after being suspended, presumably as an expression of damage of the cell surface.

We now pass to a discussion of the results obtained. This discussion concerns three virus-cell systems: the poliomyelitis virus (PV)-*rhesus* monkey kidney cells; the Newcastle disease virus (NDV)-chicken embryo cells, and the western equine encephalomyelitis virus (WEEV)-chicken embryo cells. All three viruses are known to contain RNA and presumably no DNA (Schwerdt and Schaffer, 1955; Franklin et al., 1956; Beard, 1948).

The first approach to the study of the virus growth was the determination of the "one step growth curve" (Ellis and Delbrück, 1939). In these curves the cells in suspension are uniformly infected with virus at a given time; then the excess virus is washed away and the cells are diluted in a growth medium and incubated. The amount of virus released spontaneously into the medium as a function of the time after infection is determined by the plaque assay. In this experiment the dilution of the infected cells into the growth medium is an essential step; its purpose is to decrease the probability that a cell will become reinfected with virus released during the experiment. If dense cell populations are used, extensive readsorption takes place; thus the shape of the growth curve is changed and complications may arise.

One step growth curves of several viruses revealed constant characteristics, which can be summarized as follows. The cells do not release virus for a period after infection (the latent period), then virus release begins and lasts for another period (the rise period). Finally a maximum titer is reached, which remains constant in time, or decreases, if the virus has a high rate of thermal inactivation.

If the infected cells are washed to eliminate free virus and disintegrated by physical methods which do not affect the infectivity of the virus (sonic vibration, freezing and thawing) another kind of curve is obtained. This curve describes the amount of infectious virus present within or connected with the infected cells (intracellular virus) as a function of the time. Curves of this type were obtained for WEEV (Rubin et al., 1955) and NDV (Rubin and Franklin, 1956) in our laboratory, and for PV by Howes and Melnick (1956).

In the case of WEEV and NDV it was found that the amount of intracellular virus is always less than that of the extracellular virus; thus the spontaneous release of the virus particles follows promptly after the particles have acquired the property of infectivity (maturation). It could be calculated that mature particles of these two viruses become extracellular within a few minutes; this suggests that the process of maturation of these two viruses occurs very near the cellular surface. Since in the case of WEEV it was shown that the same cell can continue to yield virus over a period of several hours (Dulbecco and Vogt, 1954), the release must involve the individual particles almost one by one, after maturation; it appears likely that this occurs through a local process of lysis of the area where virus maturation had taken place. This conclusion is in agreement with available electron microscopic evidence obtained with eastern equine encephalomyelitis virus (Bang and Gey, 1952).

In the case of poliomyelitis virus the intracellular fraction is larger than the extracellular fraction; thus this virus spends a longer time— one or two hours—within the cell after maturation. The study of yields from single cells showed that this virus, in contrast to NDV and WEEV, is released in a burst-like fashion, after having accumulated in the cell in mature form. This could be proved by determining the amount of virus released by individual infected cells as a function of the time; it was observed in these experiments that the cytoplasm of

the cell undergoes a progressive vacuolization and that the virus produced by a cell is released within a period of a half hour when the
external wall of the vacuoles breaks and their content is released into
the medium (Lwoff et al., 1955). It appears therefore that the poliomyelitis virus becomes mature relatively deep in the cytoplasm and is
set free inside cytoplasmic vacuoles.

One may speculate that the different site of maturation is related to
the different chemical composition of the viruses. NDV and WEEV
contain a large proportion of phospholipids in their external layer
(Franklin et al., 1956; Beard, 1948) whereas PV does not (Schwerdt
and Schaffer, 1955). The enzymatic systems for the synthesis of the
phospholipids may be found near the cell surface, which is known to
contain phospholipids; and this may force the NDV and WEEV to
reach their mature state near the cell surface. Similar reasoning would
apply to the influenza viruses.

We shall now briefly examine the other end of the growth curve:
the phase of the infection of the cells by the virus.

If cells infected by NDV are disrupted in the very early part of the
latent period, some of the infecting virus is still recovered in infectious
form (Rubin, 1956). This virus cannot be recovered by destroying
enzymatically the virus receptors of the intact cells. Later in the latent
period no infectious virus can be recovered by any means. It appears
therefore that the virus particles penetrate intact into the cell, and are
only later transformed into noninfectious entities. It can be speculated
that in this transformation the viral RNA is set free. This point, the
fate of the RNA of the infecting particle, is being actively pursued by
tracer methods.

Thus we have come to define a period of the infectious cycle during
which the cell does not contain any mature virus particle. This period
is defined as the "eclipse period." The essential steps of growth occur
during this period; their study is more difficult because the multiplying
entities are noninfectious and, therefore, indirect methods of study
must be applied.

Information about the character of the intracellular multiplication
could be obtained from a study of the rise period of the growth curve.
The early part of the rise period is exponential; the question whether
this exponentiality reflects an intrinsic characteristic of the process of
multiplication during the eclipse period was studied by determining
the distribution of virus yields from individual cells (Dulbecco and
Vogt, 1953). The experimental distribution was found to agree with the

theoretical expectation under the assumption that there is within the cell a pool of virus precursors which multiply exponentially and that from that pool a constant small fraction is continuously withdrawn for maturation.

Additional information about the character of the intracellular multiplication was sought by studying the loss of the ability of the monkey kidney cells infected by poliomyelitis virus to yield infectious virus as a consequence of irradiation with ultraviolet light at various times during the latent period. In this experiment, originally designed by Luria and Latarjet (1947), the number of multiplying virus precursors present within the cell at a certain time should be determined from the shape of the inactivation curve. The experiments were carried out by infecting suspensions of cells in the way described for the one step growth curve. Samples of the diluted cell suspensions were withdrawn from the growth flask at various times, and the cells were washed and suspended in a medium transparent to ultraviolet light. The cells were exposed to various doses of ultraviolet light; aliquots of the irradiated cell suspensions were then plated on cellular monolayers. Whereas the doses of ultraviolet light used did kill the cells within 24 hours, they affected to a much lesser extent the ability of the cells to yield active virus. As a control, the loss of the ability of the cells to yield active virus when irradiated before being infected was determined in a similar way.

The results did not bear out the theoretical expectations. The survival curves obtained were nearly of single-hit type at any time during the latent period. This implies that the infectious process within a cell is affected by the radiation as one or perhaps a few units at any stage of reproduction; at no stage does it behave as a complex made up of a number of independent virus precursors. It was found, in addition, that the slope of the inactivation curves decreases during the latent period to about ⅓ of that of the inactivation curve of the free virus.

This situation is very similar to that encountered in the case of the even-numbered bacteriophages (Luria and Latarjet, 1947; Benzer, 1952). The change in the slope of the curves seems to imply that changes in the state of the virus particle take place after infection. The persistence of the 1-hit character of the curves is puzzling. It may be formally explained by one of the two following, non mutually exclusive, hypotheses. The first hypothesis is that the virus becomes at a certain point integrated in a pre-existing "virus reproducing center" of the cell (Epstein, 1956) which behaves as a unit in virus multiplication; chemical models to explain this behavior could be visualized. The second hypothesis is that the infectious virus changes its state to become itself the center of virus reproduction, which constitutes the ultraviolet resistant form; this center would in turn give rise to products—virus precursors—sensitive to the radiation. The presence of the sensitive precursors would not be detected from the shape of the survival curves, if after their inactivation the ultraviolet resistant form were able to start the process of multiplication once again.

The second hypothesis could be reconciled with the notion of a geometrical multiplication of virus precursors already developed: the ultraviolet resistant form would initiate the process of multiplication, which would then proceed autonomously.

An interesting observation is that the survival curves of the intracellular virus are similar in cells infected by one or by several virus particles. This is due to another characteristic of the process of viral infection of animal cells: that in most cases only one virus particle per cell participates in the multiplication.

This information derives from experiments of mixed infection in which cell populations were exposed to two lines of poliomyelitis virus differing by one or two genetic characters. The markers used were a heat resistance marker (t) and a plaque type marker (d), to which reference will be made later. The genetic composition of the viral progeny of individual cells was analyzed. It was found that when the number of infectious particles attached to each cell was high (10–20 as average) and equal for the two lines, the majority of the cells yielded progeny constituted exclusively by one virus type. Which type appeared in the progeny depended on the genetic constitution: a temperature sensitive line (t^s) tended to exclude the temperature resistant line (t^r); and a d^- line tended to exclude a d^+ line. By carrying out experiments with unequal multiplicity of the two lines, it could be found that a cell receiving one particle of the excluding type and as many as 15 of the excluded type frequently yielded particles of the excluding type only. Under the most favorable conditions the proportion of cells yielding a mixed progeny approached 50%; in this case the two markers appeared in approximately equal proportions in the yields.

We have therefore a phenomenon of interference, whereby a cell can be infected only by one or a few virus particles. This phenomenon may be interpreted in two different ways: (1) as limitation: a cell may admit only a few particles, for instance, because it has only a few sites or centers where virus reproduction can take place. The data so far available may be compatible with the hypothesis if the number of sites or centers is in most cells equal to two. (2) As exclusion: a cell may admit only particles attaching to it within a certain time interval. The occurrence of a phenomenon of this type in animal viruses has been recently established in work carried out with NDV (Baluda, 1956): a cell infected by one or a few particles of ultraviolet inactivated NDV becomes resistant to superinfection by active NDV that attaches to the cell a few minutes later. This time dependent exclusion also occurs with poliomyelitis virus, although the time interval seems to be longer. It cannot be decided, for the time being, whether the time dependent exclusion accounts entirely for the reported results, and especially for the effect of the genetic constitution on the direction of the exclusion. Experiments intended to clarify these points are in progress.

We now turn to another aspect of our studies: the effect of the physiological state of the cells on the growth of the virus. That small changes in the state of the cells should affect the growth of the viruses in them could be anticipated from a study of the behavior of viral diseases in organisms; it is likely that in these the onset, the evolution, and the cessation of a viral disease depend in many cases on changes of the state of the cells. A spontaneously occurring mutant that supports this view was found; this is the d mutant already mentioned. This mutant produces plaques with maximum efficiency only under a narrow set of conditions (Vogt and Dulbecco, 1956); it is therefore restricted in its growth potentiality. The efficiency of plating of this mutant can be varied considerably by varying the thickness of the agar overlay, or by changing the composition of the overlay in many ways: the bicarbonate concentration—and thus the pH—of the overlay is particularly important. Under identical conditions the wild type shows its regular efficiency of plating.

Under conditions of reduced efficiency of plating, the mutant virus attaches normally to the cells and produces in them the series of changes brought about by the wild type virus (Tenenbaum, 1956); the infected cells show however a decreased virus release.

Restricted mutants were found in all three types of poliovirus. The restriction appears to persist in the animal infected by the virus, since these mutants are non-neuropathogenic in the monkey.

The basis of the restriction is unknown. It is hoped that its study may point to some process essential in virus multiplication.

In the work discussed until now the growth of the viruses was studied by analyzing physiological phenomena. It could be expected that genetic studies would open other powerful ways of attack. The following results can be presented.

We have studied the mutability of the poliomyelitis virus in respect to the two characters already mentioned, temperature resistance and restriction. Two mutation frequencies were approximately determined, $t^s \rightarrow t^r$ and $d^- \rightarrow d^+$; both were found of the order of 10^{-5}. Due to lack of selective procedures the frequency of the reverse mutations has not been determined as yet; it can be however stated that they are smaller than 10^{-2}. These data show that the virus is genetically stable. The stability in respect to the direct and reverse mutation of the same character suggests that the genetic material of the virus is not a freely segregating polyploid system. It is interesting to point out in this connection that genetic instability could be simulated by the enrichment of unfrequent mutants during the growth of the virus population, due to the interference phenomenon already discussed; one wonders, for instance, how much this phenomenon contributes to the apparent genetic instability observed in influenza virus.

Recombination studies with poliomyelitis viruses have been initiated; they were however hampered by the described selective interference. Thus this very important point cannot be discussed here.

It seems that from the described experimental results some conclusions of general nature on the comparative biological behavior of animal viruses with RNA and of viruses with DNA (bacteriophages) can be derived. It should be stressed once more that the animal virus experimentation was carried out with procedures allowing the same degree of definition and accuracy obtainable in bacteriophage work and that therefore the comparison is meaningful.

Both similarities and differences between the two systems can be detected. Which of these to emphasize is perhaps a matter of personal preference. However, it seems that the differences concern points which do not involve the most fundamental biological properties, whereas the similarities concern points which do.

The main difference between the two systems, apart from the chemical composition, is the method of penetration of viral material into the host cell and the method of release of the progeny virus from it; another difference is the absolute length of the growth cycle. All these differences can be considered as secondary, and essentially due to the different anatomy and size of the host cell in the two cases.

The similarities seem however of profound significance. Both types of viruses have an eclipse period and a similar evolution of the ultraviolet survival curves of the intracellular virus: thus both viruses undergo profound changes at the moment of penetration into the host cell or shortly thereafter. The type of multiplication is geometrical in both cases; the mutational behavior is similar, the observed frequencies of mutation having comparable values. Thus the method of multiplication and of mutation have considerable similarity in the two systems. The interference occurs with similar characteristics in the two virus types; therefore, some fundamental aspects of the virus-cell interaction are similar in the two cases.

Having stressed the importance of the similarities between the two systems, their significance should now be shown. The real meaning of the indicated similarities is for the moment unknown; however, they suggest rather definite hypotheses. Two extreme hypotheses can be visualized, such as the hypothesis that RNA and DNA are essentially similar in their basic function, or the hypothesis that RNA viruses multiply with a DNA intermediate, if not vice versa. Experimental weapons directed toward the solution of these questions are available; it is hoped that further experimental work with both virus systems will soon provide the answers.

BIBLIOGRAPHY

Bang, F. B., and G. O. Gey. 1952. Comparative susceptibility of cultured cell strains to the virus of Eastern Equine Encephalomyelitis. *Bull. Johns Hopkins Hosp. 91*, 427–461.

Baluda, M. 1956. Personal communication.

Beard, J. W. Purified animal viruses. *J. Immunol. 58*, 49–108.

Benzer, S. 1952. Resistance to ultraviolet light as an index to the reproduction of bacteriophage. *J. Bact. 63*, 59–72.

Cooper, P. D. 1955. A method for producing plaques in agar suspensions of animal cells. *Virology 1*, 397–401.

Dulbecco, R. 1952. Production of plaques in monolayer tissue cultures by single particles of an animal virus. *Proc. Nat. Acad. Sci. 38*, 747–752.

Dulbecco, R., and M. Vogt. 1953. Some problems of animal virology as studied by the plaque technique. *Cold Spring Harbor Symp. Quant. Biol. 18*, 273–279.

Dulbecco, R., and M. Vogt. 1954. One-step growth curve of western equine encephalomyelitis virus on chicken embryo cells grown *in vitro* and analysis of virus yields from single cells. *J. Exp. Med. 99*, 183–199.

Dulbecco, R., and M. Vogt. 1955. Biological properties of poliomyelitis viruses as studied by the plaque technique. *Annals N.Y. Acad. Sci. 61*, 790–800.

Dulbecco, R., and M. Vogt. 1956. Unpublished work.

Eagle, H. 1955. Personal communication.

Ellis, E. L., and M. Delbrück. 1939. The growth of bacteriophage. *J. Gen. Physiol. 22*, 365–384.

Epstein, H. T. 1956. A model of the effects of irradiation on bacterial ability to support phage growth. *Bull. Math. Biophysics 18*, 265–270.

Fogh, J., and R. O. Lund. 1955. Plaque formation of poliomyelitis viruses on human amniotic cell cultures. *Proc. Soc. Exp. Biol. Med. 90*, 80–82.

Fogh, J., and C. E. Schwerdt. 1956. Physical particle per plaque ratios observed for human poliomyelitis viruses. *Fed. Proc. 15*, 253–254.

Franklin, R., H. Rubin, and C. A. Davis. 1957. The production, purification, and properties of Newcastle Disease Virus labeled with radiophosphorus. *Virology 3*, 96–114.

Howes, D. W., and J. L. Melnick. 1956. Personal communication.

Levine, S., T. T. Puck, and B. P. Sagick. 1953. An absolute method for assay of virus hemagglutinins. *J. Exp. Med. 98*, 521–531.

Luria, S. E., and R. Latarjet. 1947. Ultraviolet irradiation of bacteriophage during intracellular growth. *J. Bact. 53*, 149.

Lwoff, A., 1956. Personal communication.

Lwoff, A., R. Dulbecco, M. Vogt, and M. Lwoff. 1955. Kinetics of the release of poliomyelitis virus from single cells. *Virology 1*, 128–139.

Melnick, J. L. 1956. Personal communication.

Rubin, H. 1956. Personal communication.

Rubin, H., M. Baluda, and J. E. Hotchin. 1955. The maturation of western equine encephalomyelitis virus and its release from chick embryo cells in suspension. *J. Exp. Med. 101*, 205–212.

Rubin, H., and R. Franklin. 1956. Personal communication.

Scherer, W. F., J. T. Syverton, and G. O. Gey. 1953. Studies on the propagation *in vitro* of poliomyelitis viruses. IV. Viral multiplication in a stable strain of human malignant epithelial cell (strain HeLa) derived from an epidermoid carcinoma of the cervix. *J. Exp. Med. 97*, 695–710.

Schwerdt, C. E., and F. L. Schaffer. 1955. Some physical and chemical properties of purified poliomyelitis virus preparations. *Ann. N.Y. Acad. Sci. 61*, 740–754.

Schwerdt, C. E., and F. L. Schaffer. 1956. Personal communication.

Tenenbaum, E., 1956. Personal communication.

Vogt, M., and R. Dulbecco. In preparation.

III. GROWTH AND DIFFERENTIATION OF PLANT TISSUE CULTURES

BY RICHARD M. KLEIN[1]

IN any discussion of cellular growth, it is incumbent upon the author to attempt to define the processes with which he is concerned. This has resulted in a plethora of definitions, not all of which are applicable to tissue cultures. It was decided, therefore, to restrict ourselves to the statement that growth of tissue cultures is the discernible manifestation of those processes resulting in irreversible increases in the numbers and sizes of cells. This discussion will be further limited by emphasizing the events or processes that control the physical alterations perceived as cell enlargement and cell division rather than on the mechanics and physico-chemical processes of enlargement and division proper.

Subsequent to the successful in vitro culture of plant parts—initially roots—several groups of workers adapted the organ culture techniques to the culture of plant tissues (White, 1954). It should be noted that tissue culture growth is a mass phenomenon in that measured growth increments represent the changes occurring in an integrated and organized mass of cells that are usually physically bound together. Caplin (1947) has diagrammed the morphological changes that occur during the growth of a solid tissue culture and has shown that growth by cell division occurs only in limited zones or "knobs" near the periphery of the culture (Fig. 1). Ball (cf. Gautheret, 1955) has discussed the idea that tissue cultures are highly organized. For these reasons, we cannot speak of plant cell cultures but, with rare exceptions (Reinert, 1956), are dealing with true tissues composed of numerous cells with a common origin, definite structure, and some differentiated or specialized members.

At this point it is necessary to delimit the various etiological categories of tissue cultures (Gautheret, 1954). The so-called "normal" tissues really represent the primary and secondary derivatives of a wound cambium (White, 1951). These are properly *callus* cultures

[1] The New York Botanical Garden. Original work discussed here was supported in part by an institutional Grant-in-Aid from the American Cancer Society, recommended by the Committee on Growth, National Research Council. Previously unreported data in this paper were obtained in collaboration with R. S. Rabin, I. L. Tenenbaum, and G. M. Groves. The critical evaluation of the paper by Drs. D. T. Klein and W. J. Robbins is gratefully acknowledged.

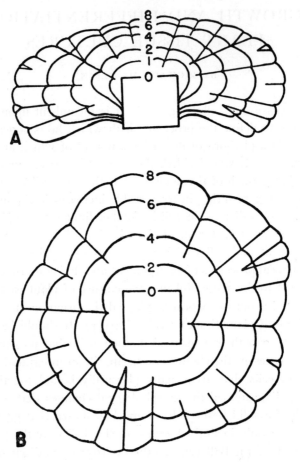

Fig. 1. Diagrammatic sections through cultures of genetic tumor tissue of tobacco illustrating morphological changes during eight weeks growth. A. Section perpendicular to surface of medium. B. Section parallel to surface of medium. Initial cube inoculum was altered into flattened hemisphere by growth of small knobs in which proliferation occurred in and near the surface. Concentric lines represent shape of culture after indicated number of weeks of growth. Radial lines indicate planes of contact between adjacent knobs. Taken from Caplin (1947) and published with the permission of the Botanical Gazette and the University of Chicago Press.

although this term has been used to designate tissue cultures from other categories. *Habituated* tissues are derived from callus cultures and have acquired the ability to grow without the addition of the growth substances required by most callus cultures (Gautheret, 1955). Crown-gall tumor tissues may be divided into two subgroups, those derived from tumors in which the bacteria were initially present (primary) and those from which the bacteria could never be isolated (secondary). The precise etiology of secondary tumors is still unknown. In a few instances, a third subgroup has been recognized—that of induced tumor tissue. Two other major categories of tissue cultures are available. One was isolated from tumors that arose spontaneously on the aerial portions of certain hybrids of tobacco (genetic tumors) and the other was derived from tumors caused by a known virus on *Rumex* and allied plant species (virus tumors) (Nickell, 1955). Demetriades (1955) has reported the culture of tissue from an insect gall and Reinert (1956) has cultured tissue from a spruce tumor of unknown etiology.

Tissue cultures from one or more of these categories have been obtained from plant species of widely separated taxonomic affinities and from many organs and tissues. Tissue cultures from the same host organ derived from etiologically distinct categories are now available. These "sets" of tissues may be grossly and microscopically indistinguishable but differences in friability, opacity, color, and rates of growth are usually observed. It is, however, usually impossible to categorize an unknown tissue visually and some prior knowledge of its etiology is essential.

I. CONDITIONS OF GROWTH

A. Nutrition. All of these tissues have many common nutritional properties, but time does not permit a review of the extensive literature on the utilization of carbon and nitrogen sources (Riker and Hildebrandt, 1954; Gautheret, 1955; Nickell, 1955). Glucose and sucrose are the best carbon sources, and nitrate is the best nitrogen source for most tissues. Somewhat surprisingly, most amino acids and other nitrogenous intermediates are either toxic or of indifferent value, although Casamino acids occasionally stimulate growth. The mineral requirements are not strikingly different among tissues nor do they deviate much from the requirements of intact plants. One notable exception to this generalization is the high phosphate requirement of tissue cultures derived from virus tumors. Painstaking studies by Heller (1953) and others have resulted in nutrient solutions that for many tissues are superior to previously devised formulae. As might be expected in studies utilizing a number of etiologically distinct tissue cultures derived from widely separated species, certain tissues have special requirements for particular ions. Trace elements, particularly iron, boron, and zinc are required for all plant cells. There will undoubtedly be additional modifications of nutrient solutions adapted to particular experimental designs but for the most part the basic nutritional patterns of tissue cultures from all etiological categories is fairly well worked out.

B. Aeration, light, and temperature. Several other facets of the conditions for cell growth have been examined. White (1953), Heller (1953), and Muir and Hildebrandt (1953) all found that liquid media, with various provisions for aeration, permitted more rapid and more uniform growth than did agar cultures. By extending the experimental period beyond the usual month or so, White found that the growth of tissues in roller tubes stopped entirely and that the slower-growing agar cultures reached greater final weights and volumes. He has suggested that this cessation of growth in liquid was not due to depletion of the nutrient, but rather that the tissues stopped growing when the replication of some cellular component or process fell behind the rate of cell division. This suggestion merits experimental confirmation with tissues of several etiological categories as well as serious consideration that, ultimately, the maximum possible rate of cell replication may be controlled by the rate of synthesis of factors or formed cytoplasmic elements that, like viruses, may be "diluted" out.

Visible radiation has received little attention in spite of the well-known morphogenic effects of light on plant growth. Green light appeared to stimulate cell division and anthocyanin production in grape callus tissue cultures (Slabecka-Szweykowska, 1955). DeCapite (1955) showed that 350 Ft.-c. of continuous white light was superior to light of lower or higher intensities for callus cultures and for secondary crown-gall tissues. When the temperature approached that optimal for growth, the effects of light were most pronounced. Hildebrandt et al. (1945) and DeCapite (1955) both noted that the optimal temperature for growth of tissue cultures is somewhat higher than for the intact plant. Henderson and Rier (1955) found that the deleterious effects of elevated temperatures on sunflower tumor tissues was relieved by casein hydrolysate, adenine, and several B-group vitamins. The obligatory participation of all of these factors was not demonstrated and the temperature-sensitive system is unknown.

II. ENERGY METABOLISM

During the past year our laboratory has been studying the relationship between cell division and the energy metabolism of the tissues. These studies are scarcely beyond the preliminary stages but the available data are sufficiently provocative to present at this time.

The growth rates of tissue cultures of the Boston Ivy, *Parthenocissus tricuspidatus* Planch., were linear for at least a month (Fig. 2). Crown-

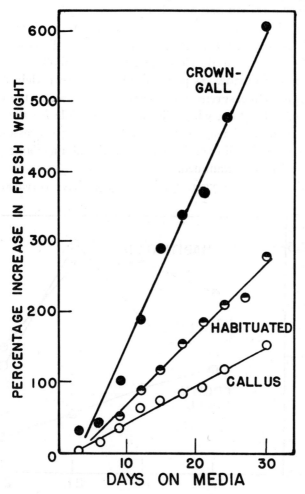

Fig. 2. Growth increments of tissue cultures of *Parthenocissus* harvested at 3-day intervals for 30 days.

gall tissue had the highest growth rate and callus the lowest of the three. Since there was no significant deviation in the percentage dry weight of these cultures during the experiment, the observed increments of fresh weight probably represent increases in cell numbers rather than increments of cell wall or cytoplasmic materials by non-dividing or enlarging cells. Tissue cultures of grape (*Vitis vinifera*), a related genus, show similar patterns of growth, as do sets of tissue from tobacco, black salsify, and others. This pattern is not absolute, for certain clones of sunflower callus have growth rates in excess of clones of crown-gall tissue (Klein and Vogel, 1956).

These data show that *Parthenocissus* tissues from etiologically distinct categories possess different potentials for cell division and, as a preliminary to precise studies on the energy metabolism, respiratory measurements were made. As pointed out elsewhere (Klein and Link, 1955), respiratory measurements in vitro rarely measure the operating capacity of cells. They do, when allowances are made for the relatively low thermodynamic efficiency of the energy-yielding processes, give us some idea of the over-all operable capacity, i.e. the metabolic potential, of the cells and tissue examined.

Data on the rates of oxygen uptake (Fig. 3) have revealed that the

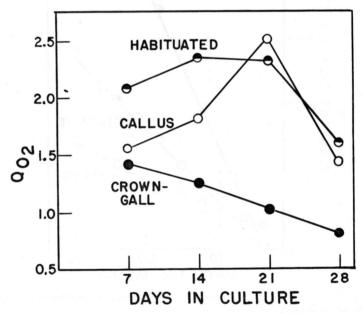

Fig. 3. Rates of oxygen uptake of slices of *Parthenocissus* tissue cultures taken for manometric study after various periods of time of growth in vitro.

rates of respiration were not constant throughout the period of study although the growth rates were constant. Further, the respiration of tumor tissue was the lowest of the three examined although its growth rate was the greatest. Since the oxidative capacities of the tissues were not in the proper relation nor of the proper magnitude to account for the utilization of energy for growth, the fermentation of these tissues was studied. A time-course study of anaerobic carbon dioxide production (Fig. 4) showed that, as for oxygen uptake, the tumor tissues

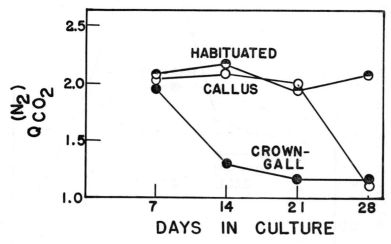

Fig. 4. Rates of fermentative carbon dioxide production of slices of *Parthenocissus* tissue cultures taken for manometric study after various periods of time of growth in vitro.

fermented at rates somewhat lower than did the other tissues. It is of interest to note that here, too, the capacity for energy formation via fermentation decreased in the tumor tissues with time. A somewhat similar decrease in the fermentation rate was noted in the callus tissues but the habituated tissues maintained their fermentative capacity during the period of study. It must be assumed, therefore, that the energy used for growth of tumor tissues relative to that used by habituated or callus tissues is either less or is stored or used more efficiently. The second of these possibilities, efficiency of storage, will be tested during contemplated studies on the oxidative phosphorylation of isolated mitochondria. Clearly, this aspect of tissue culture growth is far from solved.

Paralleling and complementing the studies on growth and respiration are examinations of the phosphorus metabolism of these etiologically distinct tissues, particularly the adenosine triphosphate metabolism. As an orienting experiment, 2,4-dinitrophenol (DNP) was tested on the growth of tissue cultures on the assumption that the DNP would inhibit growth by preventing the accumulation of the ATP necessary for cell division activities (Fig. 5). The expected response to DNP was, however, only observed with tumor tissues where there was inhibition

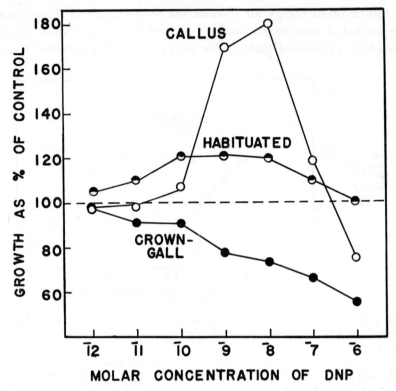

Fig. 5. Effect of 2,4-dinitrophenol on the growth of *Parthenocissus* tissue cultures. Harvested 30 days after inoculation onto White's medium (for crown-gall and habituated tissues) or White's medium containing naphthaleneacetic acid (for callus tissues).

of growth proportional to the concentration of DNP added to the medium. Habituated and callus tissues, on the other hand, responded in an anomalous manner with highly significant stimulations of growth at concentrations of DNP that are not known to affect respiratory processes or the accumulation of ATP. Identical results were obtained with a set of sunflower tissue cultures. Newcomb (1950) found that low concentrations of DNP reversibly stimulated respiration and permitted aerobic fermentation of tissues from genetic tumors of tobacco. Higher concentrations of DNP inhibited respiration and also completely suppressed aerobic fermentation. No explanation for our results is immediately forthcoming and experiments on oxidative phosphorylation of mitochondria are still to be done.

Surprisingly little information is available on any phase of the energy metabolism of tissue cultures in spite of their nearly ideal qualifications for this type of work. Cznosnowski (1952) studied the activity of several unrelated hydrolytic and oxidative enzymes from three categories of grape tissue cultures. Peroxidase activity was highest in habituated tissues and catalase was lowest, a finding confirming that of Galston (1951) on black salsify tissue. Newcomb (1951) reported vigorous ascorbic acid oxidase activity in several unrelated tissue cultures and Gentile and Naylor (1955) concluded that the oxygen uptake of virus tumor tissue was mediated almost entirely by a cytochrome system. Comparative studies were not reported. The available data on energy metabolism do not permit speculation as to whether there are modifications in intermediary metabolism resulting from tissue isolation (Plantefol and Gautheret, 1941), much less any evaluation of the alterations which undoubtedly occur as the result of habituation or the acquisition of the tumor state.

III. GROWTH SUBSTANCES

Contrasting the paucity of information on metabolic aspects of growth, the effects of growth substances has been actively investigated. In general, these studies have centered on either auxins (with indole-acetic acid [IAA] as the type compound) or on poorly characterized "growth factors." It has been assumed that increases in growth rates upon the addition of a presumed growth substance indicates that the substance in question is, or mimics the biological action of, a compound that is in suboptimal amounts or is absent from the cell or that the test substance acts to overcome the suppressive effects of some intracellular inhibitor. Failure to obtain such stimulations has all too often been taken as proof that the compound is not involved in growth.

A case in point is that of IAA in plant tumors. Although several investigators (cf. Kulescha, 1951) demonstrated that tumor tissue in vitro contain more IAA than do callus tissues, attempts to demonstrate a positive growth or metabolic response with exogenous auxin usually resulted in growth inhibitions. Proceeding on the theory that IAA is present in optimal amounts in these tissues, it was clear that only by reducing the actual or biologically active amounts of IAA could its role be demonstrated. By the use of dose levels of ionizing radiation which selectively inhibited IAA synthesis, the growth rate of tumor tissue was reduced (Table I). Near-control rates of growth could be restored by additions of IAA in concentrations which were inhibitory to tissues containing normal amounts of IAA (Klein and Vogel, 1956). As a further test of this hypothesis, a specific antimetabolite for IAA, 2,4,6-trichlorophenoxyacetic acid, was used to block its utilization.

Additions of appropriate concentrations of IAA reversed these inhibitions.

TABLE I. Effect of ionizing radiation and indoleacetic acid on growth of tissue cultures of secondary crown-gall of sunflower.[1]

Treatment	Mg Fresh Weight		Growth Increment	Std. Devia.	Per Cent Control
	Initial	Final			
Control	242	569	130	16%	100
IAA[2]	267	442	66	13%	51
Irradiated[3]	263	375	42	18%	32
Irrad. + IAA	265	545	106	14%	82

[1] Klein and Vogel (1956).
[2] IAA 5 mg/l.
[3] 1000r of x-rays given 10r/min.

A. Indoleacetic acid. The problem of auxin physiology of tissue cultures is, however, neither simple nor at the moment is it near a solution. Up to a few years ago, it was thought that all callus tissue required IAA or its biological equivalent for any growth in vitro beyond that period necessary for depletion of endogenous IAA. Habituated tissues were believed to have acquired the property of synthesizing all their auxin by enzyme induction, and crown-gall tissues synthesized adequate amounts as a result of the action of crown-gall bacteria. These assumptions have today proved to be only partially correct, and examination of some of the exceptions is instructive. Callus tissues of carrot root fall into three groups with respect to auxin. Some cultures do not require any auxin for growth (Gautheret, 1950), others require auxin (Nobecourt, 1939) and some clones will not proliferate unless other factors are supplied in addition to auxin (DeRopp et al., 1952). Gautheret (1955) has tabulated similar variations in the culture of callus from many plant species. The physiological state of the tissue at the time of isolation must also be considered. Gautheret (1955) found that callus cultures of *Parthenocissus* were easily obtained on an auxin medium only when the explants were taken from growing plants. Tissues from dormant, winter plants would not proliferate in the presence of IAA. Whether this was due to the presence of growth inhibitors active only during dormancy or is related to modifications in systems concerned with the synthesis of other required substances is unknown.

A similar spectrum of responses to auxins has been noted among habituated tissues. Gautheret (1947) first observed that when callus tissue cultures were grown in the presence of IAA for many transfers, they occasionally became capable of growth in its absence. Habituation in carrot, grape, and Boston Ivy is accompanied by increased friability, less organized cell division, and in some instances by the inability of the habituated tissues to differentiate as did the original callus (Gautheret, 1955). Nevertheless, certain strains of habituated tissue still respond positively to additions of auxin and others are indifferent or inhibited. Genetic tumors of tobacco hybrids, certain virus tumor tissues, and tissue derived from an insect gall may be grown without auxin, but their growth rates are frequently increased by the addition of one or more of these growth substances.

Fewer examples of partial or complete dependence on auxin have been reported for tissue cultures of crown-gall origin. Tumors from clover and Jimson weed are stimulated by auxin (Gautheret, 1955). Hildebrandt and coworkers (1947, 1952) reported that low concentrations of naphthaleneacetic acid stimulated the growth of secondary tumors of sunflower and primary tumors of tobacco. Much of this increased growth may have been due to excessive cell enlargement rather than to cell division. Duhamet (1955) found that when explants of habituated or crown-gall tissue of salsify weighed less than 4–10 mg their survival and growth were functions of exogenous supplies of growth substances. He proposed that these small fragments were in a negative auxin balance and required supplementation until sufficient numbers of cells were formed to permit adequate synthesis of growth substances for the culture. From his data, it appears that this critical mass is of the order to 10–30 mg. Muir et al. (1954) could only grow single isolated cells by keeping them in contact with the exudate of a growing culture.

The concentrations of auxins in habituated or crown-gall tissue cultures are remarkably constant, varying less than 15–20% over a considerable period of time (Kulescha, 1951). Although Gautheret (1955) has repeatedly suggested that habituation represents the induction of enzyme systems maintaining near-optimal levels of auxin, the sporadic nature of the alteration and the finding that usually it is only a small sector of a culture that shows this alteration had led DeRopp (1951) and Kandler (1952) to suggest that habituation is the result of a somatic mutation of one or more genes controlling the auxin-regulating system. The nondependence of genetic tumors of tobacco on auxin has been explained by Kehr (1951, 1954) on the basis that the cytoplasm of cells with combined genomes contain more endogenous auxin than that synthesized by either parent alone and that the high levels of auxin resulting from this heterosis act on cytoplasm conditioned to lower levels of auxin. A similar derangement of genetically controlled phytohormone metabolism may also be used to explain habituation. The commonly observed dependence of callus cultures on exogenous auxin and the equally widespread independence of crown-gall cultures is not surprising in view of their etiology.

B. *Coconut milk.* Several laboratories are actively investigating the chemical nature of "coconut milk factor" and, although considerable progress has been made, the exact structure of the active fraction or fractions is still unknown. First used to facilitate the in vitro growth of plant embryos, its remarkable growth-promoting properties have been exploited for the maintenance of tissue cultures. As with auxin, it now appears that the responses of different tissues to coconut milk also show considerable variation. Demetriades (1954) and Duhamet (1954) found that coconut milk significantly increased the growth rates of callus and habituated tissues from several species. It appears that the active materials are primarily cell division rather than cell enlargement factors. The same concentrations were effective on crown-gall tissues but many exceptions have been reported. Braun and Naf (1954) and Klein (1952) reported that crown-gall tissues in vitro and in vivo contain nonauxinic growth substances whose biological action is similar to coconut milk.

C. *Yeast extract.* Although numerous extracts of plant material (tomato juice, grape juice, etc.) have been reported as stimulants of cell enlargement or division in tissue cultures, yeast extract warrants special mention. This treasure chest of vitamins, amino acids, and growth factors has been used as the point of departure for the isolation of many substances. In a narrow range of concentrations, yeast extract significantly stimulated the growth of sunflower crown-gall tissues (Hildebrandt, et al., 1946) and clones of marigold crown gall tissue (Muir et al., 1955). Jagendorf and Bonner (1953) found it essential for the successful growth of cabbage tissue. Since it is not invariably effective for all tissues of any one etiological category, yeast extract, like auxins and coconut milk, cannot, a priori, be considered essential for the growth of any particular type of tissue culture.

D. *Synergistic action of growth factors.* Of particular interest are those reports on the synergistic action of yeast extract, coconut milk, and auxins. Kehr (1955) found that appropriate mixtures of the first two factors permitted the tumorous proliferation of tissues from tobacco seedlings and greatly stimulated tumor production on young stems of tumor-disposed hybrids of tobacco. The addition of auxin provoked even greater responses. Paris and Duhamet (1953) prepared a mixture of amino acids, vitamins, and auxin in proportions found in coconut milk and obtained increased growth rates of several tissue cultures, and Duhamet (1955) found that auxins, vitamins, and coconut milk synergistically increased the percentage survival and the growth of tiny fragments of various callus and tumorous tissues. Tryon (1955) obtained tumors on tobacco roots only when the tissues were supplied with mixtures of growth factors, and Tulecke (1953) found that coconut milk, yeast extract, or auxin were required for continued proliferation of *Ginkgo* tissues derived from pollen. Braun and Naf (1953) reported that their nonauxinic tumor factor was not active on tobacco pith unless auxin was added. Without auxin this factor induced only cell enlargement but was active on carrot phloem explants in the absence of auxin (Klein, 1952). Many other illustrations of the obligatory interaction of unrelated growth factors can be obtained by even a casual examination of the current literature. A nonphotosynthesizing parenchyma cell from the pith of a tobacco stem, as one striking example, apparently possesses a complete system for intermediary metabolism and for energy storage (Newcomb, 1955) but it cannot even enlarge unless an auxin is supplied (Naylor et al., 1954). For division, kinetin appears to be a second requirement (Miller and Skoog, 1955). Since the enlargement and division of populations of these cells stops entirely when the required substances are removed, these cells are auxotrophic[2] for auxins and for whatever biological intermediate or factor is mimicked by kinetin. A clone of carrot callus may be auxotrophic for only one factor, for several, or in a few instances it may be relatively or absolutely prototrophic[2] for one or all factors. Tryon's (1955) root cultures are apparently prototrophic for auxin but are auxotrophic for a yeast extract factor and Kehr's callus cultures are completely auxotrophic for a coconut milk factor and a yeast extract factor but are partially prototrophic[2] for auxin. Cultures of embryonic tissue of lupine (Lee, 1955) appear to be prototrophic for all known growth factors.

[2] Although the terms prototrophic and auxotrophic have been used in several ways, a prototrophic cell is here considered one which can synthesize from a carbon source, inorganic nitrogen, and mineral salts, the biochemical intermediate or growth factor under consideration. Auxotrophic cells lack this ability completely, and a partially prototrophic cell can synthesize sufficient amounts for some growth but growth is stimulated by exogenous supplies. Any cell may be prototrophic for one substance and auxotrophic for others.

E. Induction of growth-factor synthesis. In spite of the large numbers of examples of variable response to numerous growth factors, our knowledge of their synthesis is far from profound. It is, however, useful to attempt to rationalize the available facts and theories into a coherent generalized scheme into which the evidence relating to growth-factor synthesis by tissue cultures may be fitted (Fig. 6).

Cells in a tissue culture receive, through mitosis and cytokinesis, identical complements of genes and plasmic hereditary components which determine their upper level of metabolic and growth potential. Examining the over-all course of the synthesis of adequate amounts of any given growth factor, it is clear that a cell may be functionally auxotrophic if it is genetically deficient, possesses genes which suppress the action of necessary genes, or carries genes which act to shunt enzyme biosynthesis. For these cells to become even potentially prototrophic, mutations must occur at the points of interference as indicated in Fig. 6 by arrows. As noted earlier, the phenomenon of habituation may result from a gene mutation, the altered gene permitting the synthesis of auxin-forming systems. A known example of prototrophy being determined at this level is that of the genetic tumors of tobacco in which one parent supplies the genetic complement necessary for the synthesis of high levels of auxin.

Once the proper genetic complement is present and functional, prototrophy is dependent upon the development of enzymes and regulatory systems directing the utilization of the factor under consideration. The cytoplasmic control of potential prototrophy in cells of tissue cultures is probably more common than the genetic control since all the cells were derived from a single germ cell which, for the proper development of the organism, must have been genetically adequate for prototrophy. Further, the previously discussed findings of White (1952) and Braun (1951) that prototrophic tissue cultures may be grown at rates sufficiently rapid to "dilute" out the systems directing growth-factor prototrophy or the crown-gall state suggests that formed cytoplasmic elements are the sites of synthesis and regulation. Virus tumors may be another example of this concept.

As indicated on Fig. 6, enzyme synthesis may be completely or partially prevented by several mechanisms, and synthesis can occur only by the action of etiological agents or factors which activate, unmask, or otherwise prevent the action of cytoplasmic factors which can interfere with this synthesis. Gautheret, for example, has stated that habituation results from the activation of the necessary enzyme systems for IAA syn-

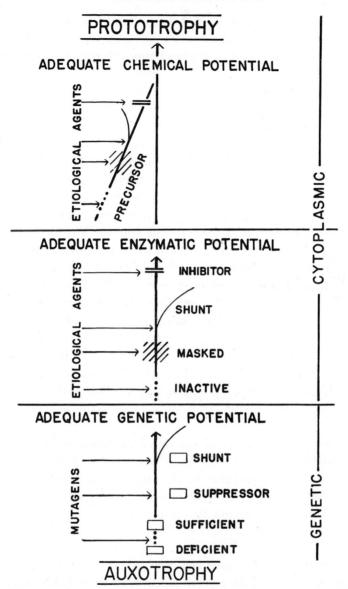

Fig. 6. Diagrammatic representation of the genetic and nongenetic factors of a cell which control the auxotrophic or prototrophic metabolism of the cell for any given growth factor. Explanation in text.

thesis rather than from a genetic alteration. Braun (1956) has presented evidence that during the process of cellular transformation to crown-gall, two growth-factor synthesizing systems are activated or unmasked. One of these is an auxin system, the other may form a factor similar to kinetin or to the coconut milk factor. The high ascorbic acid oxidase activity of tumor tissue cultures (Newcomb, 1950) may represent the activation or unmasking of this oxidase during tumor-cell transformation and the ability of crown-gall tissue cultures to utilize a wider variety of carbon and nitrogen sources than can callus tissues (Riker and Hildebrandt, 1954) may be of similar etiology. The reported presence of an inhibitor for IAA synthesis in callus cultures of sunflower and its absence from crown-gall tissues may also fit into this category (Henderson and Bonner, 1952) although this finding is unconfirmed.

Once the cell has an active and complete enzyme system for the synthesis of any growth factor, the formation of this compound is dependent upon the availability of precursors and intermediates which, in turn, may be unavailable for the same reasons. The blocking of IAA synthesis by ionizing radiation (Klein and Vogel, 1956) is an experimentally induced inhibition of the supply of a necessary precursor as would be the action in the cell of exogenous or endogenous antimetabolites. When all the necessary conditions, genetic, enzymatic, and biochemical, have been fulfilled and adequate supplies of the factor under consideration are synthesized and utilized, the cell is prototrophic for that factor. Any interruption in the synchronized and integrated process will result in a cell which is either partially prototrophic or actually auxotrophic. In the examples cited in the preceding section of this paper, these processes apparently occur for a great many factors and intermediates.

In crown-gall, the nature of several of the transforming agents is at least partially known and the time sequence of their action is well worked out. This system may be the one of choice for the elucidation of the causal action of known etiological agents on the induction of growth-substance and metabolic prototrophy.

IV. DIFFERENTIATION

We have thus far discussed cell enlargement and cell division as the complete expression of the growth potential of tissue cultures. Although these activities are the primary ones considered in tissue culture development, differentiation and maturation of cells in vitro also appears to be the result of chemical regulation. Much of this work has been done with intact plants and organs in vitro, but it has become of interest to those working with tissue cultures as well.

For our purposes, differentiation may be defined as a modification of the growth potential of cells and tissues expressed as obvious or subtle changes in gross, microscopic, or ultramicroscopic (molecular) structure or in easily or difficultly discernible changes in function. Differentiation is, however, not identifiable with the observed changes which are the visible criteria of preceding modifications in protoplasmic constitution (Waddington, 1948; Weiss, 1949). Senility will not be considered here since it is not synonymous with differentiation (Cowdry, 1956).

Parenthetically, it is impractical to assert that differentiation is or is not reversible. Reversions of plant cells from the structurally and functionally mature state to a juvenile or meristematic form are characteristic of wound-callus formation and of crown-gall transformations and the conversion of mature epidermal cells in flax hypocotyls into meristematic or bud primordia initials is an extreme case in point (Crooks, 1933). Whether these cells are differentiated (or modulated), then dedifferentiated, and then redifferentiated, or are initially potentially juvenile, or are mature at a juvenile level is purely a matter of semantic convenience.

Discussion of differentiation in plant tissue cultures can be divided into three sections dealing, respectively, with structural differentiation of cells, the differentiation of organs, and physiological differentiation.

A. Structural differentiation. When a plant cell is subjected to physico-chemical stresses which may be compensated only by alterations in its potential for development, it has only a limited range of reactability. It may respond by alterations in protoplasmic constitution and in metabolic activities which can modify cell structure, or may alter the capacity of the cell to enlarge, divide, or show morphogenic organization. It was hoped that tissue cultures would develop in an environment so stable and uniform that these stresses, unless deliberately applied, would not develop and that, under these conditions, all cells in a culture would be identical. Clearly, this hope could not be realized. Histological examination of many tissue cultures has shown that within the mass certain cells were structurally unlike their neighbors (White, 1939). Most frequently, the altered cells showed accumulations of lignin on their walls and have been called cribro-vascular elements, wound sclerids, tracheids, etc., because of their structural similarity to xylem elements in normal organs (Plate I).

A potential for sclerid differentiation seems to be acquired during the development of a callous culture. Carrot callus cultures are derived from a cambium which itself is differentiated from phloem parenchyma. Phloem parenchyma *in situ* rarely shows sclerid formation but the callus culture is highly lignified. If a sector of this callus becomes habituated, the degree of lignification decreases and, if transformed into crown-gall tissue, lignification again increases. Czosnowski (1952) found that callus tissues of grape stems contained the most lignified elements, crown-gall fewer, and habituated tissue the least of the three. Gautheret (1955) has observed similar differences in vascularization in Boston Ivy and in other species and Henderson (1954) has photographed the rather striking decrease in the number of lignified cells which are found following habituation of sunflower tissue. Not unexpectedly, different clones of tissue may vary in the degree of lignification and, occasionally, structural modifications in a clone are accompanied by changes in the amount of lignification.

The conditions of stress responsible for the modifications in potential eventuating in this type of differentiation are poorly understood. The position of the cell in the tissue mass must be considered. Restrictions of nutrients, water, and aeration are not the only parameters of this position effect. The accumulation of metabolic end products, the effects of light, and mechanical pressures require evaluation. The influence of growth substances is the only one which has received sufficient attention to permit any evaluation of its role in lignification. In an elegant series of experiments, Camus (1949) has shown that substances diffusing from buds will evoke the formation of a strand of lignified xylem elements in a parenchymatous tissue mass. These effects may be mimicked to some extent by auxins. Tryon (1955) found that differentiation of sclerids occurred rather soon after the formation of callus on tobacco roots and that these cells appeared directly adjacent to pre-existing cells of secondary xylem. Robbins et al. (1936), found that short segments of corn roots in vitro would callus but would not differentiate stelar elements unless initially long enough to contain some differentiated members. Struckmeyer et al. (1949), examined the histological effects of auxins on tissue cultures of sunflower crown-gall and found that low levels of several auxins induced significant increases in the numbers of lignified elements when compared with controls. High levels of auxin were inhibitory as compared with controls. Gautheret (1947) and Lance and Gautheret (1951) have made similar observations on callus, habituated and crown-gall tumor tissues of several species. Jacobs (1952) and Torrey (1953) have found that auxin appears to be directly implicated in the differentiation of xylem elements in intact organs.

It is still an open question whether auxins are the proximate etiological agent in lignification or whether their effects are more indirect. Siegel and Galston (1955) have suggested that an increase in peroxidase activity in plant cells is induced by peroxides formed in response to IAA applications and Siegel (1956) has demonstrated that lignin synthesis requires peroxidative oxidation of phenylpropane derivatives such as eugenol. Jensen (1955) has shown that the peroxidase activity in the prospective vascular area of roots is induced by IAA and that this induction occurs prior to morphologically detectable differentiation. Additions of conferin, a lignin precursor, to a carrot callus tissue culture increased lignification and the addition of low concentrations of IAA greatly augmented woody deposition on cell walls and induced differentiation in cells which would not otherwise have been affected (von Wacek et al., 1954). Peroxidase and peroxide were not measured.

It is, of course, not justifiable at this time to state that IAA acts as the trigger for this type of differentiation mediated by nonauxin enzyme systems. Habituated and crown-gall tissue both contain high levels of endogenous auxin but show different levels of peroxidase activity and lignification as compared with auxin-deficient callus cultures. These discrepancies may, of course, be due to concentration differences but proof is still lacking.

B. *The differentiation of organs.* Soon after the isolation of plant tissue cultures, the formation of organs such as roots and shoots was noted by several workers. Most of these observations were made on callus cultures and hybrid tumor tissues (Levine, 1937). Because of the sporadic occurrence of these organs, there was little physiological study of the factors responsible for the change. As new clones of tissue were obtained from a single isolate it became apparent that a capacity for differentiation may remain constant for many years. Nobecourt and Hustache (1954) isolated clones of callus that regularly formed roots, shoots, or both, and other clones that showed no differentiation in almost twenty years.

The first significant experiments on the factors favoring differentiation were those of White (1939). He inoculated genetic tumor tissues into an agar medium instead of placing the fragments on the surface of the medium and observed organ differentiation in normally undifferentiated tissues. White suggested that the decrease in aeration was one of the causal factors in bud initiation. Skoog (1944) reported that low light intensities, relatively low temperatures, and decreased pO_2 favored differentiation of these tissues and that the reverse conditions or the addition of auxins tended to keep the culture in the undifferentiated condition. He also noted that roots never arose prior to stem buds, a finding not applicable to other tissues which only form roots. Zarudnaya (1945) found that a low initial pH of the medium favored differentiation even though the final pH values were the same as for undifferentiated cultures started under more alkaline conditions. This finding may reflect the more rapid entrance of nutrients into cells at low pH values. Skoog and Tsui (1948) added high sugar levels, increased phosphate concentrations and the presence of adenosine to the growing list of favorable factors. Auxins were inhibitory but their suppressive effects were ameliorated by increases in the phosphate and sugar levels. Adenosine and phosphate acted synergistically.

The inhibitory effects of auxin on differentiation of many tissues have been observed. Gautheret (1955) found that 10^{-6} gm/l. of IAA stimulated rhizogenesis of carrot callus tissues but that differentiation was suppressed as the IAA concentration was increased. Levine (1947) could not suppress bud formation in differentiating clones of carrot callus with naphthaleneacetic acid but his study was rather fragmentary. Niedergang (cf. Gautheret, 1955) found that 2,3,5, triiodobenzoic acid, a reputed antiauxin, permitted bud formation on tissue cultures. Coconut milk, although containing IAA, permitted rhizogenesis in several callus tissues although auxins alone prevented differentiation of organs (Duhamet, 1955). Recently Tryon (1955) isolated several clones of root-derived callus tissues from tobacco. One clone regularly formed shoots, another has remained undifferentiated for three years, and a third reverted from differentiated to undifferentiated. Quantitative analysis of the scopoletin content of these tissue cultures showed that the undifferentiated strains were low and the differentiated strain high in this substance. Andreae (1952) has shown that scopoletin may be a competitive inhibitor for IAA oxidase and it is possible that the IAA level of the differentiated tissue remains at a level favoring differentiation of buds. On the other hand, scopoletin may have a more direct role in differentiation of organs.

The usual failure of crown-gall tissue cultures to form buds or roots (but cf. Demetriades, 1953) has excited comment. Camus and Gautheret (1948) found that bud formation on root tissue could be inhibited by grafts of crown-gall tissue and that any buds formed remained small. Habituated tissue had a similar but less pronounced effect. Crown-gall tissue cultures of periwinkle derived from tumors that had been subjected to the transformation process for suboptimal periods of time could initiate roots in the presence of auxins while those tissues receiving full transformation could not do so. Our attempts to get differentiation in normally undifferentiated crown-gall cultures by varying aeration, sugar, phosphate, etc., have thus far been unsuccessful as have attempts to obtain differentiation by control of the auxin level with an antimetabolite for IAA. Some tumor tissues, however, regularly differentiate in vivo and in vitro. These teratomas have regularly been produced on tomato, *Kalanchoe*, and tobacco plants following inoculation of certain strains of crown-gall bacteria of moderate virulence. Other strains of comparable virulence do not induce teratoma formation. Braun (1948) first isolated tissue cultures from the abnormal structures on *Kalanchoe* tumors and found that they frequently dedifferentiated into a parachymatous mass. Tobacco teratomas, however, often remain highly differentiated. Braun found that the initiation of a tobacco teratoma depended on both the strain of bacteria used and on the locus of inoculation. Undifferentiated tumors were formed in vivo following inoculation of the teratoma-inducing strain into the basal end of a cutting and differentiated tumors appeared on the apical ends. Both types of tumor formed buds and abortive shoots and leaves in vitro but when grafted back into healthy plants the initially observed morphogenic restraints were observed. Apparently three factors are involved in determining the ability of a tumor to differentiate: the strain of bacteria used, the relative position of the tumor in the host, and the inherent genetic potential of the plant cells affected.

C. Physiological differentiation. The experimental delineation of the factors composing the inherent potential of a cell in a tissue mass is much more difficult than the analogous problem in a field such as embryology. Relatively large cell populations must be used and the types of manipulation possible are correspondingly limited. The parameter of the concept most amenable to study is that of autonomy or self-determination. The acquisition of autonomy appears to be a permanent modification of cells involving their freedom from host-imposed restraints and regulations (Furth, 1953) and may occur gradually or suddenly. The bulk of this paper has been concerned with one facet of autonomy—the ability of tissue cultures to grow in vitro. Crown-gall and virus-tumor tissues express in exaggerated degree a rapid and disorganized cell proliferation and show almost complete escape from normal patterns of growth and differentiation (White and Braun, 1942). Most callus tissues require supplements of growth-regulating compounds and cannot be considered autonomous. Habituated tissues, insofar as they are prototrophic, may be autonomous.

Morphogenic autonomy, however, was more difficult to demonstrate until it was found that tissue cultures could be grafted into normal plants where their capacity to resist host-imposed patterns of growth and differentiation could be examined. The first extensive series of experiments of this type utilized heteroplastic and homoplastic grafting of secondary crown-gall tumors from sunflower (White and Braun, 1942). It was found that in successful grafts the tumor continued to proliferate in the same manner as it did in vitro instead of simply fusing with the host as did fragments of callus tissue. The chemical and physical restraints imposed by the host were incapable of inhibiting the inherent potential for morphogenic autonomy of these tumors. There was also some evidence that the "virulence" of crown-gall tumor tissues increased following grafting (White, 1945). White (1944) demonstrated the autonomy of genetic tumor tissue of tobacco hybrids by an identical procedure and DeRopp (1947) found that this grafting technique could be adapted to in vitro studies, using sterile stem fragments as the stock.

The hypothesis of Gautheret (1955) that habituation is a true tumorization was tested by the grafting technique with conflicting results. Camus and Gautheret (1948) made homoplastic grafts of callus, habituated, and crown-gall tissues of salsify onto roots of the same species. Although the callus tissue did not proliferate, both the habituated and tumor tissues formed large tissue masses. Similar results were obtained with habituated and crown-gall tissues of tobacco (Limasset and Gautheret, 1950; Lance and Gautheret, 1952) but in grape (Braun and Morel, 1950) and in sunflower (Henderson, 1954) the habituated tissue proliferated only to a slight extent and, if autonomy did exist in these habituated tissues, it was of a much lower grade than for crown-gall tissues.

Braun (1951) has shown that, in crown-gall, autonomy is acquired early in the transformation process but reaches a maximum intensity only after complete transformation. Rapidly growing tissue cultures that formed large tumors when implanted into healthy hosts required 4 days for complete transformation. Tumors which grew slowly both in vitro and following grafts were formed in only 34 hours of transformation. Autonomy for growth in vitro and morphogenic responses may also be lost. Forcing autonomous teratoma cells to divide at rates in excess of those usual in vitro by bud grafting to healthy plants resulted in a progressive loss of the tumor habit until, after several graft transfers, essentially calloid tissues were obtained in vitro which, upon grafting, acted as normal buds (Braun, 1951). Clearly, the inherent morphogenic regulators of the cells were not removed during crown-gall transformation but appear to have been suppressed or overwhelmed by the new potentials acquired during transformation.

It is evident that metabolic competence, growth factor prototrophy, and morphogenic autonomy can develop separately and each may be present in any cell or tissue with varying degrees of completeness. Under these conditions, tissue cultures show a spectrum of growth states and physiological potentials ranging from completely host-restrained through semiautonomous to the condition of complete autonomy and prototrophy characteristic of rapidly growing crown-gall or virus-tumor tissue cultures. Thus, certain callus cultures are prototrophic for required metabolites and may be metabolically competent, but they are morphogenically restrained. Others are auxotrophic for one or more growth factors but are apparently semiautonomous. Somewhat similar situations prevail for crown-gall and virus-tumor tissues although morphogenic autonomy is usually fairly well developed.

Practically nothing is known about the physiological and biochemical basis for autonomy. Undoubtedly prototrophy for growth factors, nutritional intermediates, and hormonal regulators are factors to be considered. Auxotrophic cells or tissues *in situ* can develop only so far as the supplies of necessary materials are made available to them via the vascular tissue or by polar transport. Altered capacities for energy metabolism will also affect the capacity of the cells to develop in an unrestrained manner. The structural and physiological differentiation of tissue cultures is, of course, ultimately controlled by genetic and cytoplasmic factors which can be formally treated in the same fashion as were prototrophy and auxotrophy. The capacity of a cell to proceed to structural maturity, to organize into a bud primordium, or to resist or overwhelm its original and still-present morphogenic potential must be triggered by etiological agents acting on the cytoplasmic regulatory system of the cell. The precise characterization and evaluation of all of these processes and events promise an exciting future for workers in these important disciplines.

BIBLIOGRAPHY

Andreae, W. A. 1952. Effect of scopoletin on indoleacetic acid metabolism. *Nature 170*, 58.

Braun, A. C. 1948. Studies on the origin and development of plant teratomas incited by the crown-gall bacterium. *Am. J. Bot. 35*, 511–519.

Braun, A. C. 1951. Recovery of crown-gall tumor cells. *Cancer Res. 11*, 839–844.

Braun, A. C. 1951. Cellular autonomy in crown gall. *Phytopath. 41*, 963–966.

Braun, A. C. 1953. Bacterial and host factors concerned in determining tumor morphology in crown gall. *Bot. Gaz. 114*, 363–371.

Braun, A. C. 1953. Organization of crown-gall tumor cells in the presence of a formative stimulus. *Phytopath. 43*, 204–205.

Braun, A. C. 1956. The activation of two growth-substance systems accompanying the conversion of normal to tumor cells in crown gall. *Cancer Res. 16*, 53–56.

Braun, A. C., and G. Morel. 1950. A comparison of normal, habituated, and crown-gall tumor tissue implants in the European grape. *Am. J. Bot. 37*, 499–501.

Braun, A. C., and U. Naf. 1954. A non-auxinic growth-promoting factor present in crown-gall tumor tissue. *Proc. Soc. Exp. Biol. Med. 86*, 212–214.

Camus, G. 1949. Recherches sur le rôle des bourgeons dans les phénomènes de morphogénèse. *Rev. Cytol. Biol. Veg. 11*, 1–199.

Camus, G., and R. J. Gautheret. 1948. Sur le repiquage des proliférations induites sur les fragments de racines de Scorsonère par des tissus de crown gall et des tissue ayant subi le phénomène d'accoutumance aux hetero-auxines. *C.R. Soc. Biol. 142*, 771–772.

Camus, G., and R. J. Gautheret. Sur le caractère tumorale des tissus de Scorsonère ayant subi le phénomène d'accoutumance aux hetero-auxines. *C.R. Acad. Sci. 226*, 744–745.

Cowdry, E. V. 1956. Malignant properties of cancer cells. *Ann. N.Y. Acad. Sci. 63*, 1046–1053.

Crooks, D. M. 1933. Histological and regenerative studies on the flax seedling. *Bot. Gaz. 95*, 209–239.

Czosnowski, J. 1952. Physiological features of three types of tissue of *Vitus vinifera*. *Poznan Towartz. Prayj. Nauk Prace Biol. 12*, 189–208.

DeCapite, L. 1955. Action of light and temperature on the growth of plant tissue cultures in vitro. *Am. J. Bot. 42*, 869–873.

Demetriades, S. D. 1953. Sur un cas de différenciation des tissus de crown-gall de Scorsonère cultivés in vitro. *C.R. Soc. Biol. 147*, 1713–1715.

Demetriades, S. D. 1953. Essais de culture, in vitro, des tissus des galles du *Salvia pomifera* L. *Ann. Inst. Phytopath. Benaki 7*, 61–67.

Demetriades, S. D. 1954. Action comparé du jus de tomate et du lait de coco sur trois catégories de tissus: tissus normaux, tissus de crown-gall, et tissus accoutumés à l'auxine. *Année Biol. 30*, 431–436.

DeRopp, R. S. 1947. The growth-promoting and tumefacient factors of bacteria-free crown-gall tumor tissue. *Am. J. Bot. 34*, 248–261.

DeRopp, R. S. 1951. Experimental induction and inhibition of overgrowths in plants. In *Plant Growth Substances*, ed. by F. Skoog, pp. 381–390. Univ. of Wisconsin Press, Madison.

Duhamet, L. 1955. Variations des besoins nutritifs des tissus végétaux en fonction de la taille des explantes. *Année Biol. 31*, 123–143.

Furth, J. 1953. Conditioned and autonomous neoplasms: A review. *Cancer Res. 13*, 477–492.

Galston, A. W. 1951. Sur la relation entre la croissance des cultures de tissus végétaux et leur teneur en catalase. *C.R. Acad. Sci. 232*, 1505–1507.

Gautheret, R. J. 1947. Sur les besoins en hetero-auxine des cultures de tissus de quelques végétaux. *C.R. Soc. Biol. 141*, 627–629.

Gautheret, R. J. 1950. Nouvelles recherches sur les besoins nutritifs des cultures de tissus de carotte. *C.R. Soc. Biol. 144*, 172–173.

Gautheret, R. J. 1954. Catalogue des cultures de tissus végétaux. *Rev. Gén. Bot. 61*, 672–700.

Gautheret, R. J. 1955. The nutrition of plant tissue cultures. *Ann. Rev. Plant Physiol. 6*, 433–484.

Gautheret, R. J. 1955. Sur la variabilité des propriétés physiologiques des cultures de tissus végétaux. *Rev. Gén. Bot. 62*, 1–107.

Gentile, A. C., and A. W. Naylor. 1955. The metabolism of *Rumex* virus tumors. Terminal respiratory enzymes. *Physiol. Plant. 8*, 682–690.

Heller, R. 1953. Recherches sur la nutrition minerale des tissus végétaux cultivés in vitro. Thesis, Université de Paris.

Henderson, J. H. M. 1954. The changing nutritional pattern from normal to habituated sunflower callus tissue in vitro. *Année Biol. 30*, 329–348.

Henderson, J. H. M., and J. Bonner. 1952. Auxin metabolism in normal and crown-gall tissue of sunflower. *Am. J. Bot. 39*, 444–451.

Henderson, J. H. M., and J. P. Rier, Jr. 1955. Comparative responses of normal, habituated, and tumor tissue of sunflower in vitro. *Plant Physiol. 30*, xxxv–xxxvi.

Hildebrandt, A. C., A. J. Riker, and B. M. Duggar. 1945. Growth in vitro of excised tobacco and sunflower tissue with different temperatures, hydrogen-ion concentrations and amounts of sugar. *Am. J. Bot. 32*, 357–361.

Hildebrandt, A. C., A. J. Riker, and B. M. Duggar. 1946. Influence of crown-gall bacterial products, crown-gall tissue extracts, and yeast extract on growth in vitro of excised tobacco and sunflower tissue. *Cancer Res. 6*, 368–377.

Hildebrandt, A. C., and A. J. Riker. 1947. Influence of some growth-regulating substances on sunflower and tobacco tissues in vitro. *Am. J. Bot. 34*, 421–427.

Hildebrandt, A. C., A. J. Riker, and J. L. Watertor. 1952. Growth and virus activity in tobacco tissue cultures with naphthaleneacetic acid or tryptophane. *Phytopath. 42*, 467.

Jacobs, W. P. 1952. The role of auxin in differentiation of xylem around a wound. *Am. J. Bot. 39*, 301–309.

Jagendorf, A. T., and D. M. Bonner. 1953. An atypical growth of cabbage seedling roots. III. Tissue culture and physiological comparison of typical and atypical roots. *Plant Physiol. 28*, 415–427.

Jensen, W. A. 1955. The histochemical localization of peroxidase in roots and its induction by indoleacetic acid. *Am. J. Bot. 42*, 426–432.

Kandler, O. 1952. Über eine physiologische Umstimmung von Sonnen-blumstengelgewebe durch Dauereinwirkung von β-indolylessigsaure. *Planta 40*, 346–349.

Kehr, A. E. 1951. Genetic tumors in *Nicotiana. Am. Natur. 85*, 41–64.

Kehr, A. E. 1955. Tumor induction on *Nicotiana* species by use of coconut milk and yeast extract. *Science 121*, 869–870.

Kehr, A. E., and H. H. Smith. 1954. Genetic tumors in *Nicotiana. Brookhaven Symp. Biol. 6*, 55–78.

Klein, R. M. 1952. Growth factors in crown-gall tissues. *Rept. A.I.B.S. meeting, Bot. Soc. Amer.*, Madison, Wisconsin.

Klein, R. M., and G. K. K. Link. 1955. The etiology of crown gall. *Quart. Rev. Biol. 30*, 207–277.

Klein, R. M., and H. H. Vogel, Jr. 1956. Necessity of indoleacetic acid for the duplication of crown-gall tumor cells. *Plant Physiol. 31*, 17–22.

Kulescha, Z. 1949. Recherches sur l'élaboration de substances de croissance par les cultures de tissus de vigne. *C.R. Soc. Biol. 143*, 1499–1502.

Kulescha, Z. 1951. Recherches sur l'élaboration de substances de croissance par les tissus végétaux. Thesis, University of Paris.

Lance, C., and R. J. Gautheret. 1951. Remarques sur la structure des néoformations produits sous l'action de l'acide indol-acétique sur les cultures des tissus de Topinambour. *C.R. Soc. Biol. 145*, 395–399.

Lance, C., and R. J. Gautheret. 1952. Sur la pérennité des propriétés tumorales des tissus de crown-gall et des tissus ayant subi le phénomène d'accoutumance aux auxines. *C.R. Acad. Sci. 235*, 1682–1684.

Lee, A. E. 1955. Potentially unlimited growth of lupine callus. *Bot. Gaz. 116*, 364–368.

Levine, M. 1937. Tumors of tobacco hybrids. *Am. J. Bot. 24*, 250–256.

Levine, M. 1947. Differentiation of carrot root tissues grown in vitro. *Bull. Torrey Bot. Club 74*, 312–328.

Limasset, P., and R. J. Gautheret. 1950. Sur la caractère tumorale des tissus de tabac ayant subi le phénomène d'accoutumance aux hetero-auxines. *C.R. Acad. Sci. 230*, 2043–2045.

Miller, C., and F. Skoog. 1955. Regulation of growth in tobacco tissue cultures. *Plant Physiol. 30*, xxxv.

Muir, W. H., and A. C. Hildebrandt. 1953. Growth of tissue cultures under several conditions of aeration. *Rept. A.I.B.S. meeting, Am. Soc. Plant Physiol.*, Madison, Wisconsin.

Muir, W. H., A. C. Hildebrandt, and A. J. Riker. 1954. Plant tissue cultures produced from single isolated cells. *Science 119*, 877–878.

Muir, W. H., A. C. Hildebrandt, and A. J. Riker. 1955. Concentrations of yeast extract and growth rates of marigold tissue in liquid culture. *Plant Physiol. 30*, xxxvi.

or, J., G. Sander, and F. Skoog. 1954. Mitosis and cell enlargement without cell division in excised tobacco pith tissue. *Physiol. Plant. 7*, –29.

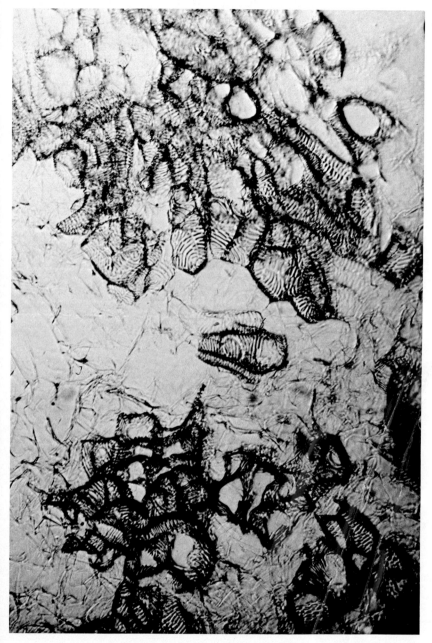

PLATE I. Photomicrograph of a section from a tissue culture of secondary crown-gall of sunflower showing sclerids. (Henderson, 1955.)

Newcomb, E. H. 1950. Tobacco callus respiration and its response to 2,4-dinitrophenol. *Am. J. Bot. 37*, 264–271.

Newcomb, E. H. 1955. The use of culture tissue in a study of the metabolism controlling cell enlargement. *Année Biol. 31*, 195–214.

Nickell, L. G. 1950. Effect of certain plant hormones and colchicine on the growth and respiration of virus tumor tissue from *Rumex acetosa. Am. J. Bot. 37*, 829–835.

Nickell, L. G. 1955. Nutrition of pathological tissues caused by plant viruses. *Année Biol. 31*, 107–121.

Nobecourt, P. 1939. Sur la pérennité et l'augmentation de volume des cultures de tissus végétaux. *C.R. Soc. Biol. 130*, 1270–1271.

Nobecourt, P., and G. Hustache. 1954. Evolution des caractères morphologiques et anatomiques dans des cultures de tissus végétaux. *Cong. Int'l Bot. 8(11)*, 192–193.

Paris, D., and L. Duhamet. 1953. Action d'un mélange d'acides amines et de vitamines sur la prolifération des cultures de tissus de crown gall de Scorsonère; comparaison avec l'action du lait de coco. *C.R. Acad. Sci. 236*, 1690–1692.

Plantefol, L., and R. J. Gautheret. 1941. Sur l'intensité des échanges respiratoires des tissus végétaux en culture; tissu primitif et tissu néoforme. *C.R. Acad. Sci. 123*, 627–629.

Reinert, J. 1956. Dissociation of cultures from *Picea glauca* into small tissue fragments and single cells. *Science 123*, 457–458.

Riker, A. J., and A. C. Hildebrandt. 1954. The carbon and nitrogen nutrition of gall tissue in culture. *Année Biol. 30*, 283–296.

Robbins, W. J., M. Bartley, and V. B. White. 1936. Growth of fragments of excised root tips. *Bot. Gaz. 97*, 554–579.

Siegel, S. M. 1956. The chemistry and physiology of lignin formation. *Quart. Rev. Biol. 31*, 1–18.

Siegel, S. M., and A. W. Galston. 1955. Peroxide genesis in plant tissues and its relation to indoleacetic acid destruction. *Arch. Biochem. Biophys. 54*, 102–113.

Skoog, F. 1944. Growth and organ formation in tobacco tissue cultures. *Am. J. Bot. 31*, 19–24.

Skoog, F., and C. Tsui. 1948. Chemical control of growth and bud formation in tobacco stem segments and callus cultures in vitro. *Am. J. Bot. 35*, 782–787.

Slabecka-Szweykowska, A. 1955. On the influence of the wave length of light on the biogenesis of anthocyanin pigment in the *Vitis vinifera* tissue in vitro. *Acta Soc. Bot. Polon. 24*, 4–11.

Torrey, J. G. 1953. The effect of certain metabolic inhibitors on vascular tissue differentiation in isolated pea roots. *Am. J. Bot. 40*, 525–533.

Tryon, K. 1955. Root tumors on *Nicotiana affinis* seedlings grown in vitro on a malt and yeast extract medium. *Am. J. Bot. 42*, 604–611.

Tryon, K. 1956. Scopoletin in differentiating and nondifferentiating cultures of tobacco tissue. *Science 123*, 590.

Tulecke, W. R. 1953. A tissue derived from the pollen of *Ginkgo biloba*. *Science 117*, 599–600.

von Wacek, A., O. Hartel, and S. Meralla. 1954. Über den Einfluss von Coniferinzusatz auf die Verholzung von Karottengewebe bei Kultur in vitro. *Holzforsch. 2/3*, 58–62.

Waddington, C. H. 1948. The genetic control of development. *Symp. Soc. Exp. Biol. 2*, 145–154.

Weiss, P. 1949. Problems of cellular differentiation. *Proc. Nat'l Cancer Confer. 1*, 50–60.

White, P. R. 1939. Potentially unlimited growth of an excised plant callus in an artificial media. *Am. J. Bot. 26*, 59–64.

White, P. R. 1939. Controlled differentiation in a plant tissue culture. *Bull. Torrey Bot. Club 66*, 507–513.

White, P. R. 1944. Transplantation of plant tumors of genetic origin. *Cancer Res. 4*, 791–794.

White, P. R. 1945. Metastic (graft) tumors of bacteria-free crown-gall on *Vinca rosea*. *Am. J. Bot. 32*, 237–241.

White, P. R. 1951. The role of growth substances in vegetative development as exemplified in tissue cultures. In *Plant Growth Substances*, ed. by F. Skoog, pp. 247–252. Univ. of Wisconsin Press, Madison.

White, P. R. 1953. A comparison of certain procedures for the maintenance of plant tissue cultures. *Am. J. Bot. 40*, 517–524.

White, P. R., and A. C. Braun. 1942. A cancerous neoplasm of plants. Autonomous, bacteria-free crown-gall tissue. *Cancer Res. 2*, 597–617.

Wiggans, S. C. 1954. Growth and organ formation in callus tissue derived from *Daucus carota*. *Am. J. Bot. 41*, 321–326.

Zarudnaya, K. 1945. The effects of *p*H on growth of tobacco callus tissue cultures. Thesis, Johns Hopkins University.

IV. RELATIONS BETWEEN CELL GROWTH AND CELL DIVISION

BY DAVID M. PRESCOTT[1]

T HERE are various levels at which the study of growth may be examined. In a multicellular organism composed of a heterogeneous population of various growing and nongrowing cells, the usual concern is with the organism as a whole, and relatively little information about the contribution of individual cells or groups of cells is obtainable. In the direction of less complicated situations, cultures of cells, preferably clonal, represent a considerably simplified multicellular situation which may yield reliable information, at least about the average cell, during the different phases of mass culture growth. A culture of a microorganism such as *Tetrahymena*, for example, follows the typical sigmoidal growth curve for a cell culture, as illustrated in Fig. 1. After inoculation into fresh nutrient medium, actively growing cells which are increasing in number exponentially, cease division for a period of

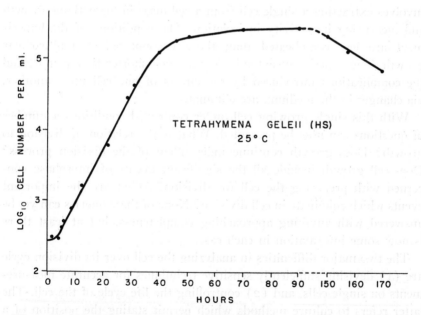

Fig. 1. The population growth of the ciliated protozoan, *Tetrahymena geleii*, on a medium of 1% proteose peptone broth. The growth curve, with 4 main phases, is typical for cells grown in mass culture.

[1] Department of Anatomy, School of Medicine, University of California, Los Angeles.

time referred to as the lag phase. This is followed by a phase of logarithmic growth with a constant average generation time, a stationary phase of constant cell density, and a final phase of decline in the population.

The lag phase appears to represent a period of readjustment by the cells to the fresh medium without significant alteration of the medium (Prescott, 1956a). The dynamic biochemical basis for the lag phase is unknown, but it may well consist of readjustments in enzyme levels and in the strengths of particular synthetic pathways. In this sense, the lag period in culture growth may be related to the lag period observed in experiments designed to demonstrate specific enzyme adaptation (see, for example, Stanier, 1954). Cell division is inhibited until the readjustments are completed.

In any event, the phases of culture growth are defined in terms of cell density, and derived information is largely statistical in nature. The principal concern is with change in the cell number, and for the most part, the particulars of the growth-duplication processes by which each individual cell progresses from division to division are only secondarily a part of the measurements. This brings us to the third level of approach to the study of growth, the single cell and the single division cycle. This involves extracting a single cell from a cell mass in logarithmic growth and measuring its activity in isolation. The conditions of the experiment are less complicated than those encountered in multicellular growth. The experiment extends over a much shorter time course, and the complications introduced by the effects of one cell upon another, via changes in the medium, are eliminated.

With this single growing cell under controlled conditions a number of questions can now be presented. What is the relation of division to growth? Does growth continue independent of the division process? Does cell growth include all the significant events of interphase concerned with preparing the cell for division? What are the important events which culminate in cell division? None of these queries can yet be answered with anything approaching completeness, but at least there is now some information in each case.

The two major difficulties in analyzing the cell over its division cycle are (1) devising sufficiently sensitive techniques for accurate measurements on single cells, and (2) controlling the life cycle of the cell. The latter refers to culture methods which permit stating the position of a cell in time with respect to mitosis. The recent rapid advances in the field of cell nutrition have greatly reduced the seriousness of this problem.

As an experimental cell type, many of the protozoans fit admirably because of their large size and ease of controlled culture. Large size, of course, permits more accuracy in measurements. There are a number of studies on growth in volume of individual protozoans, but the majority of these is incomplete and of limited interest. From this earlier work there is an indication that growth in volume is most rapid following cell division (Popoff, 1908; Chalkley, 1931). Volume changes are not necessarily always accurate reflections of changes in the metabolic machinery and whenever possible should be checked with other types of measurement. Accurate estimates of dry mass of single living cells, for example, may eventually be possible with the interference microscope.

I. GROWTH STUDIES ON SINGLE AMOEBAE

Some of my own work has centered around the growth of one particular cell, *Amoeba proteus*, whose large size makes it convenient for studies on single cell growth. Culture methods developed by James (1953) permit control of its life cycle. Thanks to the Cartesian diver balance, an instrument devised by Zeuthen (1948) in Copenhagen, it has been possible to weigh single living amoebae with great accuracy (Prescott, 1955). With sufficient patience one can obtain such precise division to division growth curves as are illustrated in Fig. 2. Each curve commences with the reduced weight (weight when immersed in water) of a daughter cell (about 10 millimicrograms) as it comes from a cell division, and each point thereafter represents the increase in weight with time. The amoebae act as their own controls in these experiments, since under the culture conditions provided, we know that the next division will occur in 24 to 25 hours. The error involved in any particular weight measurement is less than 1% (Prescott, 1955).

These data illustrate several points which are worthy of emphasis. The variation of daughter cell weights (9 to 11 millimicrograms) and the growth curves generally fall into a narrow range. Since the completion of these studies, a better understanding and control of environmental conditions for these organisms have been achieved, and the variation in daughter cell size can be reduced to a much smaller range than indicated in Fig. 2. This is another clear case of "biological variation" stemming from the experimenter's inability to provide a standard environment. "Biological variation" might well be replaced, in many situations at least, by the expression "biologist's variation." Rate of growth in amoeba rises to its highest point immediately following cyto-

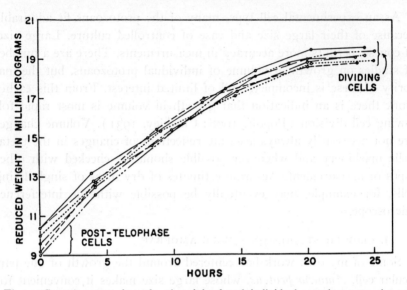

Fig. 2. Growth curves in reduced weight for 6 individual amoebae over the cell division cycle at 23°C. Each curve begins with the reduced weight of a daughter amoeba and ends with the reduced weight of the same amoeba as it enters division.

plasmic fission, falls off slightly during most of the interphase, and ceases completely as the daughter cell weight reaches the doubled value (20 millimicrograms) at about 20 hours. During the last several hours of the cycle no growth is detectable, and the cell gives no indication of impending mitosis. The division cycle consists of a period of growth lasting 20 hours, a predivision period of about 4 hours representing an unknown, and division itself lasting about 30 minutes. Most simply stated, it appears that completion of growth does not act as an immediate or direct trigger to cell division, and some events other than growth control the length of interphase and the initiation of mitosis. The absence of growth during division confirms earlier work, particularly that of Zeuthen (1951) on the ciliate, *Tetrahymena*. In addition, there is a suggestion of a predivision period of no growth in *Tetrahymena*, but it is relatively shorter.

Not all of the attempts to measure amoeba growth were successful. For example, there were cases of amoebae in which, for reasons still unknown, division was delayed to a point far beyond the normal 24 to 25 hours. Two such cases appear in Fig. 3. It is important to note that (1) these cells with long generation times are of normal size, and (2) the increase in generation time occurs mainly as a part of the predivision period of nongrowth rather than as an extension of the growth period.

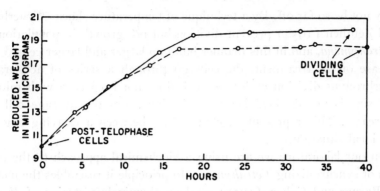

Fig. 3. Division to division growth curves for two individual amoebae which failed to divide on the normal 24-hour schedule (23°C). The predivision period between the completion of growth and the onset of mitosis is greatly extended.

In the next division cycle these "slow cells" return to the usual 24- to 25-hour generation time. These experiments serve to emphasize the existence of a predivision period.

II. SYNCHRONOUS CELL POPULATIONS FOR THE ANALYSIS OF THE DIVISION CYCLE

Rather than to rely on weight as the sole criterion of growth, procedures were devised for measuring increase in total cell protein and volume, again as a function of the division cycle. Unfortunately, measurements on single cells, even large ones, are difficult and probably always will be. In particular, biochemical analyses of the division cycle are desirable, but as yet microchemical methods are insufficiently sensitive to allow much progress in this direction. The obvious alternative to the tedious and often impossible task of single cell measurements consists in working with populations of cells synchronized in growth and division.

The controlled fertilization and subsequent cleavage of eggs represents a case of excellent division synchrony of large masses of cells. The analyses of Mazia (1955) on the biochemistry of the mitotic apparatus illustrate how profitably this type of material can be used for certain problems of the division cycle. However, growth in a segmenting egg is difficult to assess, because the egg contains large amounts of stored raw materials and no mass changes are involved. Ideally, we need a population of cells such as amoebae, which double in size between divisions. Techniques for synchronizing the divisions and growth of microorganisms are developing rapidly, and attainment of the ideal system

[63]

may not be so far off. By a technique of temperature shocks, Scherbaum and Zeuthen (1954) prevent division but not growth in populations of *Tetrahymena*, and in this manner build up larger and larger cells. Upon release of the treatment, the cells go through a series of accelerated, synchronous division cycles; growth does not keep pace with mitosis, cell size is not doubled between divisions, and the average cell size decreases. This represents a compromise between a cleaving egg and the ideal situation.

In our laboratory we are using a biochemical approach to the problem of synchronizing *Tetrahymena*. In principle it resembles the method of Barner and Cohen (1955) used on thymineless strains of *E. coli*. *Tetrahymena* are grown on the complete synthetic medium (Elliott et al., 1954), but the medium is divided into two fractions, the first containing all the required amino acids and growth factors, and the second containing only the nucleic acid components. By maintaining a mass of cells alternately in one medium and then the other, a high degree of synchrony is induced with no evident change in average cell size. The divisions take place in the nucleic acid-components medium. The method is only in preliminary stages of study; the initial results are very encouraging. Probably the alternation of media results in some degree of separation in time between protein synthesis and nucleic acid synthesis, and this underlies the synchrony. Incorporation of labeled amino acids and nitrogen bases should help to reveal the extent of the separation in synthesis.

James (1953) has succeeded in maintaining small clones of amoebae in synchrony through the use of a temperature cycling between 18° and 26°C, as opposed to the temperature shock method of Scherbaum and Zeuthen on *Tetrahymena*. In asynchronous mass cultures of amoebae, even a single shift in temperature from 18° to 26°C is enough to give a marked increase in the number of dividing forms. These can be collected to form a synchronous group of several hundred cells. With such groups increase in total protein, cell volume, and nuclear volume could be followed as a function of the division cycle. The curve in Fig. 4 represents growth in protein; each point was determined by a colorimetric measurement (Lowry et al., 1951) on 60 amoebae fixed at that point of the cycle. Although not so precisely defined, this curve is identical to the growth by weight curves for individuals. The predivision period is clearly present. Growth in cell and nuclear volumes were determined in a similar manner (see Prescott, 1955 for methods); amoebae were flattened down to a thickness of 6 microns in a compression chamber

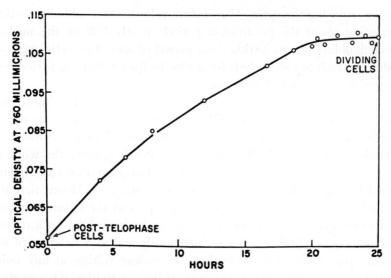

Fig. 4. A growth curve in protein for amoeba over the complete division cycle. Each point represents a colorimetric determination of protein on 60 amoebae fixed at a particular time of the division cycle.

and nuclear and cytoplasmic volumes computed by measuring the area presented by the nucleus and cytoplasm. Cell volume follows the same pattern as cell protein and weight (see Fig. 5). This indicates that any of these three criteria (weight, protein, and volume) are suitable for growth measurement in this particular cell. Apparently cell density does

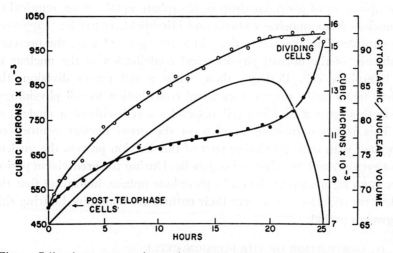

Fig. 5. Cell volume, ⊙ ; nuclear volume, ● ; and the cytoplasmic nuclear volume ration for amoeba from division to division. Each point is the average of measurements on 14 to 22 amoebae.

not change during the cycle. The nuclear volume curve adds interesting information about the predivision period; nearly half of the nuclear volume doubling occurs in this short period of no cell growth, suggesting that we look to the nucleus for a clue to the predivision period and to the initiation of cell division as well.

What events does the nuclear volume curve reflect? We know that a major portion of the amino acid turnover in amoeba proteins occurs either in the nucleus itself, or is directly and immediately controlled by the nucleus (Mazia and Prescott, 1955). Further, from the work of James (1954) and the ingenious nuclear transplantation experiments of Goldstein and Plaut (1956) comes very strong evidence that the nucleus is the point of synthesis of at least part of the ribonucleic acid found in the cytoplasm and nucleus. One or both of these nuclear contributions may well take place discontinuously and as a function of the division cycle. It is conceivable that the nuclear volume at any point may reflect the intensity of one or both of these activities. The question of how such activities might participate in preparing the cell for division would still need an answer.

III. PHOSPHATE UPTAKE AND THE DIVISION CYCLE

We have one further point of information on the division cycle. Phosphate uptake by amoebae is proportional to the cell volume during interphase, including the predivision period. During division itself, however, phosphate uptake drops by 70% (Mazia and Prescott, 1954). The prior discovery of a similar drop in phosphate uptake upon removal of the nucleus from amoebae (Mazia and Hirshfield, 1950) has suggested that phosphate uptake may reflect, in a very general way, the presence or absence of the normal physiological contribution of the nucleus to the interphase cell. Perhaps, then, when a cell enters division, the nucleus temporarily suspends its usual contribution to cell physiology, and in this sense a dividing cell resembles a cell without a nucleus at all. This is not surprising in view of the gross nuclear contortions associated with mitosis; during part of the division process the nucleus is not even present as a discrete organelle. During the predivision period there is no significant variation in phosphate uptake, indicating that the nuclear contributions, whatever their nature, are maintained during this nongrowth period.

IV. COMPOSITION OF THE DIVISION CYCLE

With the knowledge gained from these experiments we are now in a position to give at least partial answers to some of the questions posed earlier concerning cell growth-division relationships. Not only does growth cease during division, but its completion is separated from mitosis by several hours. This separation in time suggests in turn that perhaps the link between growth and division is not really a close one. What is the basis for this separation? As a first guess, we might consider the division cycle to be composed of two concurrently advancing processes, (1) growth, which results in a *doubling in cell size*, and (2) a biochemical differentiation, the completion of which brings the cell to the division state and is thus responsible for a *doubling in cell number*. In amoeba, growth is completed several hours ahead of biochemical differentiation. For the present we can speak of biochemical differentiation only in vague terms, but there is increasing evidence that the concept is a valid one. The division cycle is not occupied by a smooth, homogeneous duplication of the whole cell. Rather, particular processes are limited to particular time intervals within the cycle. The nuclear growth curve probably reflects such synthetic discontinuities. There is now good evidence that DNA synthesis is restricted to a relatively short part of the interphase (Pelc and Howard, 1952; Walker, 1953). DNA synthesis with either concurrent or subsequent chromosome doubling may be the important event we are seeking as the termination of biochemical differentiation and the turning point towards mitosis.

We must not ignore the fact that the mechanics of division entail considerable rearrangements within the cell. We have only such visible criteria as dissolution of the nuclear membrane, condensation of the chromosomes, etc., to signify that division is imminent; exactly how far the preparations extend back into interphase is unknown. Swann's (1954) studies on energy-cell division relationships and the biochemical investigations of Mazia (1955) on the structure and formation of the mitotic apparatus have clearly demonstrated the necessity of including the interphase cell in a study of cell division biochemistry.

V. THE STIMULUS FOR CELL DIVISION

The often discussed hypothesis that completion of growth triggers division is appealing in its apparent simplicity. This general proposition has been expressed more specifically in terms of a critical cell volume-nuclear volume ratio or a critical cell surface-volume ratio which might act as a division trigger. Expressed in another manner, the single nucleus is considered capable of supporting a defined maximum of cytoplasm and no more. Attainment of the maximum would mark the end of growth and stimulate in an unknown manner nuclear and cytoplasmic fissions. This idea of limited capacity of the nucleus for supporting cell growth apparently stems principally from observations on cells containing two or more nuclei or a single nucleus of a higher-than-usual ploidy. The size of such cells is usually increased roughly in proportion to the increase in the genetic material. This is certainly true for binucleated amoebae, and other investigators have demonstrated the relationship in a wide variety of cells, for example, Gerrasimov (1902) and Fankhauser (1945).

The amoeba growth curves demonstrate that attainment of a particular cell mass does not act as the immediate stimulus for division. Moreover, the separation in time between growth completion and division presents us with two problems where previously there appeared to be only one. If size stimulated cell division, then division in turn would act as a control on final cell size. The lack of a close relationship between size and division necessitates that we now explain (1) the stimulus to cell division, and (2) the control of cell size. The first question has already received a brief consideration above in terms of biochemical differentiation over the division cycle. With respect to control of cell size, we do have some information; the greater cell size associated with increased sets of genetic material has already been mentioned. The growth studies on individual amoebae show that each cell almost exactly doubles in weight over the cycle, and the division size for all cells is very nearly the same. This indicates that the control of cell size is rather rigid.

There is an additional experiment which bears on this problem of size control. Ordinarily, amoebae divide into two daughter cells of very nearly equal size. If amoebae are exposed to strong light or gently shaken during division, the subsequent splitting is often very unequal (Prescott, 1956b). Thus, in the next division cycle one cell starts off with an extra quota of cytoplasm, while the second is deficient in this material. What effects do these two situations have on growth rate, generation time, and the size attained at the next division? Figs. 6 and 7 answer these questions. The weight growth curves II and III in Fig. 6 are for two sister cells with a 67% size difference between them at birth. Curve I describes the growth course for a normal-sized daughter cell. The large cell grows more slowly than a normal cell, but still reaches division slightly ahead of the 24-hour schedule. It does not double its own size but rather grows only to the size normal for this cell at division. The small daughter cell grows rapidly but not swiftly enough to divide on schedule. In attaining the normal division size it has far more than doubled its initial weight. The weight of a cell at division, then, is

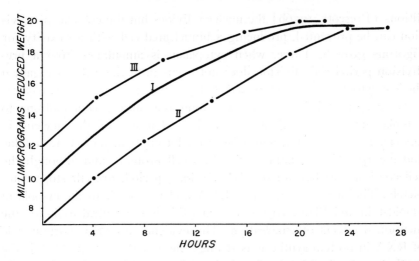

Fig. 6. Division to division growth curves in reduced weight for three daughter amoebae with different initial sizes. Curve I, growth of a cell of normal size; Curves II and III, growth of an abnormally small and an abnormally large daughter cell, respectively.

rigidly controlled and does not depend on size at birth. The experiments cited thus far indicate a genetic control of cell size, but the actual mechanisms remain to be explained. It should be noted that altering the size of a cell at birth does not have much effect on the length of the predivision period.

VI. CONTROL OF CELL GROWTH

Control of cell size implies control of the growth process itself, and the problem is perhaps more properly stated in terms of growth rather than size. We are not interested in cell size as much as in what it may reveal about the process whereby a cell grows. The general form of all the growth curves (Figs. 2, 3 and 6) is the same. The rate of growth drops gradually as mass increases. This inverse relation carries the suggestion of an auto-inhibition of growth, but there is little that can be said about this possibility in view of the lack of information. Let us consider the older proposition that a cell continues to grow until it contains the maximum amount of cytoplasm which the single nucleus can support. What is the evidence? Growth in predivision stage amoebae can be re-initiated in three ways, each of which involves increasing the amount of nuclear material per amount of cytoplasm. One, in the course of normal events growth begins immediately following cell division and reconstitution of the two daughter nuclei. Two, under certain conditions (Prescott, 1956b) the nucleus divides but the cytoplasmic division can be prevented, producing a binucleated cell with a capacity for vigorous growth. Three, when cytoplasm is amputated from a predivision period cell, division does not occur until the cell has regained the lost cytoplasm through growth (Hartmann, 1928; Prescott, 1956c).

More specific information concerning the nuclear contributions to cytoplasmic processes is now available. Removal of the nucleus in amoeba results in immediate effects on locomotion, phosphate uptake, and incorporation of amino acids into cell proteins, and although the cell continues to live for a relatively long period, growth stops completely. The nucleus synthesizes RNA and releases it to the cytoplasm (Goldstein and Plaut, 1956). Perhaps this is the central basis for the mentioned effects of enucleation. The evidence for an important role of RNA in protein synthesis is strong; the hypothesis that RNA is the carrier of genetic specificity from the chromosomes to points of protein synthesis in the cytoplasm is receiving much attention (Gale and Folkes, 1954).

In any case, let us suppose that the nucleus is operating at a constant and maximum rate in the production of some factor which is essential for growth and is used up in the process. Suppose also that the growth rate depends on the dominance of synthesis over breakdown; protein turnover data (Mazia and Prescott, 1955) indicate that the cell is in a continual state of flux. The rate of synthesis could be considered constant, since it would be in turn dependent upon a nuclear activity of constant rate. The rate of breakdown is probably a function of the amount of cytoplasm. This is supported by experiments which show that the rate of loss in weight of starving amoebae is roughly proportional to amount of cytoplasm remaining (Holter and Zeuthen, 1948; Prescott, 1956d). Under these conditions the net growth rate would decrease with increase in cytoplasmic mass, and growth would cease when breakdown equaled synthesis (during the predivision period). Further growth would depend on an increase in the rate of nuclear contribution through increase in functional nuclear machinery.

VII. CELL SIZE AND GENERATION TIME

Although the arguments presented here tend to discourage consideration of growth completion as a stimulus to division, still the information in Fig. 6 indicates that generation time is at least partially dependent upon daughter cell size. The relationship between generation time and daughter cell size is described by the data in Fig. 7. The larger the

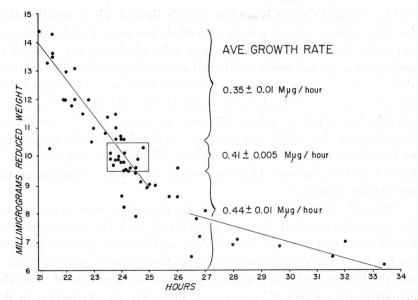

Fig. 7. Each point in the graph represents the reduced weight of a single daughter amoeba plotted against the subsequent generation time for this same amoeba. The abnormally small (less than 9.5 millimicrograms) and the abnormally large (greater than 10.5 millimicrograms) daughter cells were produced through unequal divisions of amoeba exposed to strong light.

daughter cell, the sooner it divides and vice versa. Let us take the theoretical case of a daughter cell provided with the full amount of cytoplasm normally contained in a dividing cell. By extrapolation of the upper left part of the curve in Fig. 7 we can make a rough calculation of the generation time for such a cell; it would be about 16 hours, even though this daughter cell would be of usual division size and need not grow to divide. Such "theoretical" cells have been produced by removing one nucleus from a newly formed binucleated cell, and the generation time is usually between 13 and 18 hours, reasonably close to the calculated value. These cells show only slight growth, at most. The 13 to 18 hours is perhaps the time required to complete the biochemical differentiation of the division cycle for a cell of this abnormally large size; it could represent, for example, the time required for DNA synthesis and chromosome doubling.

In all the amoeba experiments presented thus far a common factor has been the constant cell size at division (20.0 ± 0.1 millimicrograms; standard deviation ±0.7 millimicrograms). The question arises whether this size is a requirement for division, and the answer is clearly no. Amoebae in which size is limited by controlled starvation will also divide, provided the size is not too severely limited. The generation time for these cells is always greatly extended, however, and division may not occur for several days. Daughter cells which are not allowed to undergo any growth never divide and eventually cytolyze. The minimal size requirement for division has not yet been precisely defined, but amoebae must complete *at least* one-third of the full division to division growth, i.e. weigh 13 micromilligrams or more, in order to divide. Growing amoebae which are repeatedly reduced in size by amputation of cytoplasm continue to grow without dividing, as long as size is limited in this manner. Division occurs soon after cessation of the amputation treatment.

Although division is not immediately dependent upon size, perhaps the rate of biochemical differentiation is; through this route, size may have its effect on generation time. We could imagine, for example, that the biochemical differentiation of the nucleus governs generation time, and the biochemical differentiation processes in turn depend upon the cytoplasm for sources of energy, raw materials, etc. Reduction in the amount of cytoplasm could in this way slow biochemical differentiation and increase the generation time. The short generation time for abnormally large daughter cells may be the acceleration effect on bio-chemical differentiation events of an extra abundance of cytoplasm.

Summing up very briefly: the separation in time between the completion of cell growth and the onset of cell division indicates that the events of the division cycle fall into two classes: (1) growth and (2) biochemical differentiation. The first is concerned with the doubling of cell size, and the latter terminates in cell division and controls the increase in cell number. They proceed concurrently. The control of both classes of events probably resides within the nucleus. The two processes are interdependent, but the link between them is by no means a rigid one. A further testing of these hypotheses will require more information concerning the details of nuclear-cytoplasmic interrelationships and the progressive biochemical events which make up the division cycle.

BIBLIOGRAPHY

Barner, H. D., and S. S. Cohen. 1955. Synchronization of division of a thymineless mutant of *Escherichia coli*. *Fed. Proc. 14*, 177.

Chalkley, H. W. 1931. The chemistry of cell division. II. The relation between cell growth and division in *Amoeba proteus*. *Publ. Hlth. Rep. 46*, 1736–1754.

Elliott, A. M., L. E. Brownell, and J. A. Gross. 1954. The use of *Tetrahymena* to evaluate the effects of gamma radiation on essential nutrients. *J. Protozool. 1*, 193–199.

Fankhauser, G. 1945. The effects of changes in chromosome number on amphibian development. *Quart. Rev. Biol. 20*, 20–78.

Gale, E. F., and J. P. Folkes. 1954. Effect of nucleic acids on protein synthesis and amino acid incorporation in disrupted staphylococeal cells. *Nature 173*, 1223–1227.

Gerrasimov, J. J. 1902. Die Abhängigkeit der Grosse der Zelle von der Menge ihrer Kernmasse. *Z. allg. Physiol. 1*, 220–258.

Goldstein, L., and W. Plaut. 1955. Direct evidence for nuclear synthesis of cytoplasmic ribose nucleic acid. *Proc. Natl. Acad. Sci. 41*, 874–880.

Hartmann, M. 1928. Ueber experimentelle Unsterblichkeit von Protozoen-individuen. Ersatz der Fortpflanzung von *Amoeba proteus* durch fortgesetzte Regeneration. *Zool. Jb. 45*, 973–987.

Holter, H., and E. Zeuthen. 1948. Metabolism and reduced weight in starving *Chaos chaos*. *Compt. rendus des Travaux Lab. Carlsberg. 26*, 277–296.

James, T. W. 1953. Quantitative studies on the ribonucleic acid content in *Amoeba proteus* in relation to the nucleus and the division cycle. Ph.D. Thesis, Univ. of California, Berkeley.

James, T. W. 1954. The role of the nucleus in the maintenance of ribonucleic acid in *Amoeba proteus. Biochim. Biophys. Acta 15*, 367–371.

Lowry, O. H., N. J. Rosebrough, A. L. Farr, and R. J. Randall. 1951. Protein measurement with the Folin phenol reagent. *J. Biol. Chem. 193*, 265–275.

Mazia, D. 1955. The organization of the mitotic nucleus. *Symposia of the Society for Exptl. Biol. 9*, 335–357.

Mazia, D., and H. I. Hirschfield. 1950. The nucleus-dependence of P^{32} uptake by the cell. *Science 112*, 297–299.

Mazia, D., and D. M. Prescott. 1954. Nuclear function and mitosis. *Science 120*, 120–122.

Mazia, D., and D. M. Prescott. 1955. The role of the nucleus in protein synthesis in *Amoeba. Biochim. Biophys. Acta 17*, 23–34.

Pelc, S. P., and A. Howard. 1952. Chromosome metabolism as shown by autoradiographs. *Exptl. Cell Res. 2* (suppl.), 269–278.

Popoff, M. 1908. Experimentelle Zellstudien I. *Arch. Zellforsch. 1*, 245–379.

Prescott, D. M. 1955. Relations between cell growth and cell division. I. Reduced weight, cell volume, protein content, and nuclear volume of *Amoeba proteus* from division to division. *Exptl. Cell Res. 9*, 328–337.

Prescott, D. M. 1956a. Change in the physiological state of a cell population as a function of culture growth and age. *Exptl. Cell Res.* (Submitted for publication.)

Prescott, D. M. 1956b. Relations between cell growth and cell division. II. The effect of cell size on cell growth rate and generation time in *Amoeba proteus. Exptl. Cell Res. 11*, 86–94.

Prescott, D. M. 1956c. Relations between cell growth and cell division. III. Changes in nuclear volume and growth rate and prevention of cell division in *Amoeba proteus* resulting from cytoplasmic amputation. *Exptl. Cell Res. 11*, 94–98.

Prescott, D. M. 1956d. Unpublished experiments.

Scherbaum, O., and E. Zeuthen. 1954. Induction of synchronous division in mass cultures of *Tetrahymena piriformis*. *Exptl. Cell Res. 6*, 221–227.

Stanier, R. Y. 1954. The plasticity of enzymatic patterns in microbial cells. In *Aspects of Synthesis and Order in Growth*, ed. by D. Rudnick, Princeton University Press, Princeton.

Swann, M. M. 1954. The control of cell division. In *Recent Developments in Cell Physiology*, ed. by Kitching. Butterworths Scientific Publications, London.

Walker, P. M. B. 1953. Interphase synthesis of DNA in tissue culture cells. *Heredity 6* (suppl.), 275–283.

Zeuthen, E. 1948. A Cartesian diver balance weighing reduced weights (R.W.) with an accuracy of ± 0.01 micrograms. *Compt. rendus des Travaux Lab. Carlsberg 26*, 243–266.

Zeuthen, E. 1951. Segmentation, nuclear growth and cytoplasmic storage in eggs of echinoderms and amphibia. *Pubbl. Staz. Zool. Napoli 23* (suppl.), 47–69.

V. AN OSCILLATOR MODEL FOR
BIOLOGICAL CLOCKS

BY COLIN S. PITTENDRIGH AND VICTOR G. BRUCE[1]

I. THE UBIQUITY OF BIOLOGICAL CLOCKS

THE generalization we are concerned with in this paper is that most organisms, perhaps all, can measure time; that they possess clocks as part of their total adaptive organization. This generalization is based on fairly recent developments in spite of the fact that much of the pertinent evidence is comparatively old. This anomaly is itself interesting—a reminder of how a change in viewpoint can transform a problem. Thus the generalization—as such—leans heavily on the fact that an older literature reveals persistent daily and lunar rhythms to be extremely widespread phenomena; but it had to await recognition of such rhythms as manifestations of an underlying time measurement. It is this point of view that is of relatively recent origin.

We have attempted to trace the growth of the new outlook and notice that explicit use of the word "clock" is a development that largely follows the brilliant work of Kramer (1952; see also Matthews, 1955) and his associates at Wilhelmshaven on the time-compensated sun navigation of birds. Kramer's is not the first clear demonstration of chronometry by an organism; Wahl (1932) and Stein-Beling (1935) had shown in the early 1930's that the bee possesses what they called a time sense (*Zeitgedächtniss*). But it seems to us that it was Kramer's experiments which served to reorient thought about persistent rhythms, and to initiate their present treatment as a reflection of functional clocks.

The change of view implied in recognizing rhythms as clocks is immediately reflected in the type of question to which the experimenter has paid attention in the last ten years. Where "rhythm" is a strictly descriptive word, "clock" is loaded with a functional connotation: it implies a device to measure time. And in this way it sets off a line of thought which "rhythm" fails to evoke. The line of thought concerns the functional prerequisites a clock must satisfy: it demands that the rate of the clock's time-measuring "motion" be independent of major environmental variables, like temperature, which are open to wide

[1] Both authors are in the Department of Biology, Princeton University. This work was supported by a grant from the National Science Foundation and by funds from the Eugene Higgins Trust allocated to Princeton University.

fluctuation; it demands that the clock be susceptible to synchronization with the cyclic external change to which it is functionally related. It is interesting in considering the historical growth of the clock problem to contrast discussions on the bee's *Zeitgedächtniss* with those on persistent rhythms twenty years ago. Wahl and Stein-Beling, well aware they were dealing with a time measurement, directed their attention to the severe problems posed by the functional prerequisites; and they were thus led to the first discovery of the temperature-independent feature. With one exception (that of Kalmus who also worked on the bee) the contemporary workers on persistent daily rhythms do not seem to have considered the functional issues. We are familiar with only two other discoveries of temperature independence in the study of rhythms prior to 1954 and the newer outlook we date from Kramer. In neither case (Welsh, 1938; Brown and Webb, 1948) is there any indication that the effect had functional significance in reflecting the operation of a "clock." Prior to 1954 most workers were, in fact (and in consequence?), commonly concerned with explicit demonstration of temperature *dependence*. This feature, overestimated and misinterpreted, has been a prime argument in favor of the endogenous or metabolic nature of persistent rhythms. A potentially confusing issue is involved here hinging on the equivocal meanings of temperature dependence and independence. Rhythms are certainly not temperature-insensitive, but their steady-state period is virtually constant over a wide range of temperatures, and to this extent it measures time, *qua* period, in a temperature-independent fashion like a mechanical clock. The *sensitivity* of rhythms to temperature is manifest in their entrainability to temperature cycles and their development of transients (see below) following single temperature stimulations of the step or pulse type. The occurrence of such transients has evidently been mistaken earlier for a dependence of the steady-state period on temperature.

Following the change of view towards rhythms, attention has been focussed largely on temperature effects, resulting in the striking discovery that temperature independence of period is an extremely widespread phenomena now known in the bee (Wahl, 1932), *Cambarus* (Welsh, 1938), *Uca* (Brown and Webb, 1948), *Pilobolus* (Uebelmesser, 1954), *Avena* (Ball and Dyke, 1954), *Mytilus* (Rao, 1954), *Drosophila* (Pittendrigh, 1954), *Oedogonium* (Bühnemann, 1955b), *Talitrus* and *Talorchestia* (Pardi and Grassi, 1955), *Paramecium* (Ehret, 1956), *Gonyaulax* (Hastings and Sweeney, 1956), *Euglena* (Bruce and Pittendrigh, 1956), *Phaseolus* (Leinweber, 1956; Bünning and Leinweber, 1956), *Peromyscus* and the bat (Rawson, 1956), and in *Neurospora* (Pittendrigh, Bruce, Rosensweig, and Rubin, 1957). There is, in fact, no case of a persistent daily rhythm known to us which has been shown critically to have a steady-state period that is dependent on the temperature regime beyond a trivial factor (see discussion following Pittendrigh, 1957).

Discovery of widespread temperature independence has, on the one hand, been stimulated by the view that rhythms are manifestations of innate clocks and, on the other, it has provided this view with its strongest empirical support. Temperature independence is certainly not a feature one anticipates of a randomly chosen metabolic system and its universal association with rhythms invites or even demands the view that one must invoke natural selection to account for such an intuitively improbable organization.

Even a brief history of their treatment as clocks would be incomplete without reference to the repeated inference that persistent rhythms are learned phenomena. We believe that this learning approach has served to divert attention to the wider problem of a model for memory. As we note later, Kalmus (1938) clearly approached our own line of thought on the mechanism of rhythms but later discarded it for a learning model. The recent writings of Harker (1953, 1956) and Thorpe (1956) show that learning ideas are still a force in the field.

A descriptive classification of biological clocks has been given earlier (Pittendrigh, 1957), in which three categories are distinguished: (1) continuously consulted clocks (true chronometers), (2) interval timers, and (3) pure rhythms. These distinctions relate, of course, only to the functional significance (known or presumed) of what we later call the *overt persistent rhythm*; and may, or may not, reflect differences in the ultimate clock mechanism involved in each case.*

Continuously consulted clocks are now known in birds (Kramer, 1952; Hoffmann, 1954; Rawson, 1956), in amphipods (Pardi and Papi, 1953; Pardi and Grassi, 1955), in bees (Wahl, 1932; Stein-Beling, 1935; von Frisch, 1950), and apparently in the isopod *Tylos* (Pardi, 1954). Their characteristic feature is that they yield information to the organism that identifies any point in the day relative to dawn.

* *Added in Proof*: Rawson, K. S. (Sun Compass Orientation and Endogenous Activity Rhythms of the Starling (*Sturnus vulgaris* L.), *Zeit. f. Tierpsychol.*, *11*, 446-452, 1954) makes this point clearly by choosing the easily studied activity rhythm of the starling for an experimental approach to the nature of the chronometer used in this species' sun navigation.

Interval timers are exemplified by the host of insect eclosion rhythms where the temporal distribution of activity occurs at discrete intervals timed by an endogenous device; see e.g. Pittendrigh (1954), Harker (1953), Bateman (1955), and innumerable others, the most outstanding of which are summarized in tabular form by Bruce and Pittendrigh (1957). The sufficient mechanism necessary for the interval timer is formally much simpler than that of continuously consulted clocks; information is given on only one point in the cycle.

Pure rhythms are exemplified by Brown's extensive studies (Brown et al., 1955) of the daily change in the color of *Uca*; and by the daily rhythm of phototaxis in *Euglena* (Pohl, 1948; Bruce and Pittendrigh, 1956). The functional significance of pure rhythms is, it so happens, generally much less obvious than that of interval timers or continuously consulted clocks. Yet their possession of all the strongest functional prerequisites of a timing device leads us to include them as a provisionally distinct class of biological clocks.

Pure rhythms merit two further comments at this point. First, it may well be that the particular overt persistent rhythm which the experimenter utilizes may itself have little functional meaning; it may be quite incidentally controlled by a clock mechanism utilized for other purposes. In the case of the *Uca* color rhythm, for instance, it occurs to us that its remarkable precision may reflect utilization of the basic clock mechanism for solar navigation of the type exemplified by *Tylos* and still to be uncovered in *Uca*.

Second, our conviction that the temperature-independent, phase-labile rhythms themselves are—or reflect—a functional time measurement, leads us directly to a general model of the clock as an oscillator on which we elaborate in the subsequent sections.

II. A WORKING HYPOTHESIS AND A COMPARATIVE APPROACH

As several recent writers (e.g. Tinbergen, 1953; Simpson, Pittendrigh, and Tiffany, 1957) have pointed out in explicit terms, the biologist has three lines of explanation open to him: causal, functional, and historical. These are not so much alternative modes of explanation as essential complements. The living system is never fully explained, even in logical principle, by any one of them alone; and the fact, attested now by abundant experience, is that progress in any one line of explanation is often dependent on insights derived from another. The history of the study of rhythms is as good a case as any where lack of interest in functional explanation was in part responsible for slow progress in the discovery of the system's most interesting features, and incidentally those features of greatest interest to the student pursuing causal explanation.

The problem of biological clocks is full of interest in all three aspects, but in this paper we are concerned primarily with the attempt to develop a causal explanation; we want to know what the clock mechanism is. Our constant reference to functional and historical considerations is therefore prompted by our belief that they could well prove valuable guides.

Our working hypothesis, which is wholly consistent with the available facts, assumes that all organisms are capable of time measurement; that their clocks have a common and ancient basic mechanism like the hair spring and balance wheel of diverse human timepieces from wrist watches to clocks; and that this basic element is an oscillatory system with a natural period evolved to match, approximately, the period of those environmental variables that are ecologically significant (day, month, etc.). The oscillator basis of the clock is suggested directly by *overt persistent rhythms* and, as later shown, is supported by much experimental detail.

Our adoption of an hypothesis as broad as this needs both justification and qualification. In the first place, we find it both attractive and plausible in its own right. The continuously consulted clock of higher metazoa must be an extremely elaborate system, an evolutionary exploitation of a less complex precursor which afforded the opportunity for development of a complete chronometer. And it simply seems likely to us that the more common interval timers and pure rhythms are representative of that earlier precursor. The notion that the basic element is indeed ancient is supported by our own demonstration that a protist rhythm satisfies all the functional prerequisites of the higher clocks (Bruce and Pittendrigh, 1956; Pittendrigh, 1957). The ancient nature of the clock in mammals is to be inferred from Rawson's (1956) fine demonstration that the rhythms of mice and bats are temperature-independent even when the animal's homoiothermy is destroyed experimentally. We take this to mean that the mammal clock must be at least as old as the Paleocene and probably much older. One cannot invoke selection to establish an autonomously temperature-independent system in a homoiotherm.

The historical outlook, however, itself raises a *caveat* to our optimism. It is commonplace that selection is indifferent to the precise form in which a functional problem is solved; and the consequence is that an identical adaptive problem is met in diverse organisms by a multiplicity of different solutions. It may be noted, however, that the principle of multiple solutions usually applies to cases where the same problem has been met in different phyletic lines independently of each other. The likelihood of encountering different causal mechanisms behind a common functional activity is greatly reduced when the function is itself ancient and demands a mechanism that is, in some sense, improbable. In the latter case the innate conservatism of organization will render it likely that a working solution once attained will be retained.

Apart from its intrinsic attractions the working hypothesis has three further merits associated with the comparative study it suggests. First, to the extent it is sustained by further work, it permits the experimenter free choice of material in asking a particular question; there are many clock problems best tackled in flies and others for which *Neurospora* or *Euglena* are more suited. Second, it permits us to tackle questions concerning the level of biological organization necessary for a clock. Discussion of the third and by far the most important merit is deferred to a later section where we argue that the comparative approach is itself an essential tool in the systems analysis, necessary to disentangle effects attributable to the common basic mechanism from those attributable to the peripheral physiological systems the clock controls.

In general it seems unnecessary now to review the evidence which convinces us that in considering clocks one is dealing with an integral feature of the living system and not with its *forced* response to an external periodicity. Pittendrigh (1957) has recently discussed the problem and notes that the most powerful evidence concerns the natural period of persistent daily rhythms which in *virtually all* cases we know of is not exactly that of the solar day. It should, however, be mentioned that one worker in the field (Brown et al., 1956) apparently retains some skepticism on the point.

In the following section we develop and use a formal model for the mechanism of biological clocks based on an endogenous self-sustaining oscillation (ESSO).

III. THE DEVELOPMENT AND USE OF A DESCRIPTIVE OSCILLATOR MODEL

A. The terminology and characteristic features of periodic systems. We should emphasize at the outset that what we have in mind by the term formal model is really only a consistent verbal framework for the description and interpretation of existing data and phenomena. Of course, our speculations are influenced by concrete examples of oscillators, but our aim is, insofar as possible, to present a descriptive terminology for the discussion of periodic systems which will be useful without committing us immediately to a physico-chemical or physiological mechanism. This model, like any other, serves the dual functions of facilitating the interpretation of existing data and phenomena and of guiding further speculation and experimentation.

Fig. 1 is a block diagram showing in schematic form the organization of the biological clock system and its relation to environmental variables.

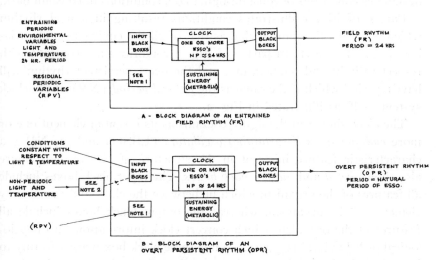

Fig. 1. Block diagrams of the clock-oscillator model. Two cases are shown. A. Conditions resulting in a field rhythm (FR) in which the period will be equal to the period of the environmental variables. B. Conditions constant with respect to light and temperature resulting in an overt persistent rhythm (OPR) with a period determined by the inherent properties of the clock. The clock includes as a basic element one or more endogenous self-sustaining oscillators (ESSO). For further description, see the text.

Note 1: All periodically fluctuating environmental variables other than light and temperature are called residual periodic variables (RPV). They are assumed to be never coupled to the oscillator.

Note 2: Light shocks, temperature shocks, and non-periodic treatment with light and temperature may affect the phase of OPR.

Information about the system is derived from observation of the various features and properties of the experimentally observed rhythm which may concern locomotion, respiration, color change, eclosion, phototaxis, or any other overt biological activity. We call these *overt persistent rhythms* (OPR) if they persist in the laboratory under conditions constant with respect to light and temperature. They are distinguished from *field rhythms* (FR) which occur in the presence of light and temperature cycles. It is important to note that the definition of persistent rhythms recognizes the continuance of periodicity in many other variables besides the excluded light and temperature. These *residual periodic variables* (RPV) include pressure, humidity, air ionization, cosmic ray showers, and doubtless others. The term *constant conditions* refers only to the light and temperature regime. It is customary in referring to cycles of alternating light and dark to use the terminology "LD 12:*12* (8–20)" for a periodic cycle of 12 hours of light: *12* hours of dark, in which dawn is at 8 a.m. and sunset at 20 (or 8 p.m.) LL designates conditions of constant light; DD, conditions of constant dark.

Part A of Fig. 1 illustrates conditions resulting in a field rhythm which will have the same period as that of all the environmental variables (e.g. 24 hours). In part B of Fig. 1 conditions are constant with respect to light and temperature, and the overt persistent rhythm will have a period which is the same as the *natural period* (NP) of the clock system itself, as illustrated in Fig. 2.

The *clock* shown in the figure contains as its essential element one or more *endogenous self-sustaining oscillators* (ESSO) and the NP is a characteristic feature inherent in the structure of ESSO. The figure also shows *input* and *output black boxes*. The input black boxes include all features of the organism which lie between the clock and the sensory elements of the organism, whereas the output black boxes include all features of the organism which convert clock information into physiological or behavioral responses. While these black boxes are a feature to be reckoned with by the experimenter, they cannot be expected to have common properties in different organisms, and we shall make no attempt to characterize them more concretely. We do, however, wish to characterize ESSO a little better both by way of explaining our selection of a self-sustaining oscillator as the basic element of the clock, and by enumerating the characteristic features of such oscillators.

We should, however, first point out that although the OPR of an individual may persist indefinitely there are some rhythms (notably those involving a population of organisms) which do not. This apparent damping of the rhythm, treated briefly in a later section, and in footnote 2, p. 83, is included within the scope of the model we are discussing here. One other feature of the block diagram should be mentioned before we discuss the characteristics of ESSO. What we have labelled "clock" may differ in details from one organism to another. Thus, the complete clock system in the higher metazoa undoubtedly has evolved features not found in the protists. We do, however, feel that there is a basic

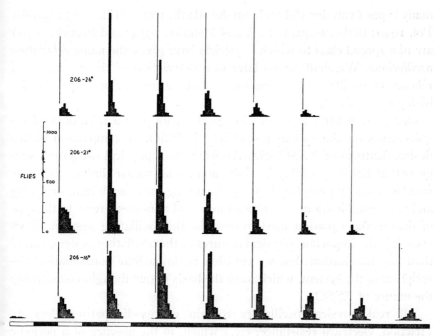

Fig. 2. Illustration of an overt persistent rhythm (OPR) in *Drosophila* eclosion obtained in constant darkness at 26° (top), 21° (middle), and 16° (lower). A light:dark (LD) cycle of a 24-hour period established the phase of the rhythm. Vertical guide lines are 24 hours apart and ordinates represent numbers of flies eclosed per hour. Compare the variance on the peaks in the entrained OPR (this Fig.) with the variance in the "initiated" OPR (Fig. 3). Also, note the slight influence of low temperature in lengthening the period.

ESSO with an NP close to 24 hours which satisfies the functional prerequisites and is common to all the clocks. In a later section of the paper we speculate concerning ways in which the more complex chronometers might have evolved from the fundamental ESSO to achieve features not ordinarily associated with the clock systems found in the simpler organisms.

The self-sustaining feature is the outstanding characteristic of ESSO which suggests the type of oscillator model we are looking for.[2] It is a property characteristic of biological rhythms in general and this feature, as well as others, has led previous workers to suggest that heart beats, nerve discharges, and in particular short-period biological rhythms of

---

[2] One could also construct an oscillator model in which the apparent damping of rhythms would be attributed to actual damping of the oscillator itself. Such a model could account for some of the characteristic features of endogenous rhythms, but would be unsatisfactory for the interpretation of indefinitely persisting rhythms with periods other than exactly 24 hours.

many types (van der Pol and van der Mark, 1929; Hill, 1933; van der Pol, 1940; Bethe, 1940; Franck and Meunier, 1953; and Franck, 1956) are of a special class to which physicists have given the name *relaxation oscillations*. We shall return later to a description of those features of relaxation oscillations which make them attractive as a model for biological rhythms.

Our present knowledge of biological clocks puts a minimum of requirements on the formal properties of ESSO. A comparison of the desired features of ESSO with the features of physical oscillators suggests that ESSO probably is of the non-linear, rather than of the more familiar linear type exemplified by the tuning fork, pendulum, or spring and mass combination. The terms linear and non-linear refer to the type of differential equation used to describe the oscillating system. However, it is the important physical features of the oscillating system, rather than any implication that we envision an immediate mathematical description of the system, which form the basis of our thoughts concerning the nature of ESSO.

All real physical oscillators contain energy-dissipating elements (friction, electrical resistance, etc.) which, in the absence of a periodic energy input, may cause the oscillations to damp out and disappear. Some physical oscillators contain in addition energy-absorbing elements to which the term *negative damping* has been applied. An oscillator will be *self-excited* and *self-sustaining* if, for small amplitude oscillations the net damping is negative and for large amplitude oscillations the net damping is positive. The system will approach a steady-state oscillation where the energy absorbed just equals the energy dissipated in each cycle. Such an oscillation is excited and persists if the energy is supplied to the oscillator at a constant rate; and the natural period of the oscillation is characteristic only of the structure of the oscillator. A wide variety of physical (mechanical, electronic, etc.) oscillators of this general type are known. We do not wish to imply, however, that all self-sustaining oscillations must be dependent on negative damping nor that they must be self-excited. However, in the interest of being specific we would like to single out oscillators of the above type, and in particular a special class of such oscillators, as being specially appropriate for the model of ESSO.

A feature of oscillators which we have so far not mentioned is their wave form. In self-sustaining oscillators it may be sinusoidal, as it is in the linear cases, but it may be of almost any other form. Relaxation oscillators characteristically generate a wave form showing very abrupt changes. It is characteristic of them that the system periodically reaches an unstable condition which is followed by a more or less discontinuous jump to another state and the period of the oscillation is determined by some form of *relaxation time* required for the system to reach the unstable condition.

There are several very characteristic features of relaxation oscillations other than their self-excitation and self-sustenance. These relate to the steady-state oscillations arising as a result of periodic energy inputs or as a result of the coupling of oscillators. The period of a *forced* relaxation oscillation, i.e. one resulting when the system has a periodic energy input, will not necessarily be the period of the forcing agent. It is only when the period of the forcing agent is sufficiently close to the natural period of the relaxation oscillation that this will be the case, and then the relaxation oscillation becomes *entrained* by the periodic external agent. The difference in period between the OPR and the FR in Fig. 1 is due to the fact that in the latter case the environmental periodicities of light and temperature are coupled to ESSO and *entrain* it (see below) to the period of a solar day. RPV's are not energetically coupled to ESSO and hence fail to entrain it with the result that OPR's reflect the inherent natural period of ESSO.

If two similar oscillators of nearly the same frequency are coupled together (allowing energy transfer between them), then it may happen that they *mutually entrain* each other to some intermediate frequency. Mutual entrainment may also occur if the two oscillators have quite different frequencies; then the mutual entrainment results in each oscillator completing a different whole number of cycles in a given time interval. A related phenomenon, called *frequency demultiplication,* may result when a periodic energy input into an oscillator has a frequency which is roughly an integral multiple of the frequency of the oscillator. Frequency demultiplication occurs if the resulting oscillation is entrained to a whole number submultiple of the high frequency oscillation.

The features of entrainment and frequency demultiplication are characteristic of steady-state oscillations and, except in their detailed quantitative aspects, they are generalizations which one can make without being very specific about the detailed formal properties of the relaxation oscillation. The detailed formal properties might include considerations such as the general nature of the wave form of the oscillator (saw-tooth wave shape, square wave, etc.), or the correlation of the phase of the oscillator with the phase of OPR and the phase of the entraining agency.

In principle it should be possible to investigate some of these formal properties of the oscillator by examining both the quantitative aspects of entrainment and by examining the response of the oscillator to non-periodic disturbances. Such responses, when they are not periodic, are called *transients*. Generalizations about the transient response of relaxation oscillators are difficult to make; nevertheless, for specific types of oscillators and specific types of disturbances, some generalizations are possible. In the following parts of this section we utilize the oscillator model to interpret the major features of existing data. We have taken the opportunity afforded us by the invitation to speculate freely in this symposium!

B. *Exclusion of the learning interpretation.* Persistent rhythms have long been subject to interpretation as learned phenomena; the adjective *persistent* itself predisposes one to the idea that OPR's are a learned hang-over from a FR somehow involving memory of the FR's period. This view goes back at least as far as Semon (1912) although it is here complicated by a Lamarckian blurring of the distinction between (true) ontogenetic learning and phylogenetic "learning."[3] Semon's writings evidently influenced Kalmus (1940b) whose papers (1933, 1935, 1937-1938, 1938, 1940a, and 1940b) are among the most important in the rhythm literature of the last thirty years. In one of his reviews (1938) Kalmus gives clear evidence of beginning the same line of thought we develop here. Thus he makes an explicit pendulum analogy for persistent rhythms and introduces the term *eigenfrequenz* which he uses in a manner nearly identical to our own use of *natural period*. The full implications of this language were, however, not developed by Kalmus. In his last paper on the *Drosophila* eclosion rhythm Kalmus (1940b) abandoned it in favor of the Semon memory terminology. Here he develops a gramophone record model for the mechanism of persistent rhythms. This record is supposed to be "cut" in the individual's own history by experiencing the 24-hour interval between successive dawns; and it is replayed as the memory store (*of period*) on later occasions when the organism is prevented from seeing the light regime. This particular model, incidentally, was also conceived to accommodate Kalmus' belief that the *Drosophila* clock was basically temperature

[3] The evolution of clocks is a subject full of interesting problems which we deliberately avoid in the present discussion devoted to the question of clock mechanism. Perhaps more clearly than any other feature of the organism, its OPR's invite a Lamarckian interpretation; so much so that Bünning (1956) still finds it desirable to comment that a selectionist interpretation is possible. Pittendrigh (1957) briefly comments on its probable evolution through "organic selection."

dependent. Thus the original "record" was cut while spinning at a particular speed determined by the prevailing temperature; and on being replayed gave an overestimate of period at lower temperatures and an underestimate at higher temperatures. It is in discussing this model that Kalmus resorts to the Semon terminology of engrams, ekphory, etc. It is unnecessary to stress further the extent to which this abandons and contradicts the earlier ideas (Kalmus, 1935, 1938) implied in his use of *eigenfrequenz*.

The interpretation of OPR's as truly learned is evidently still favored by some workers. Thorpe's (1956) book on instinct and learning discusses OPR's as cases of imprinting; and Harker (1953, 1956), whose outstanding work on the cockroach we discuss later, also speaks of OPR's as "impressed rhythms." We believe the experimental evidence to be in such strong contradiction to this view and the issue at stake so fundamental for the field as to demand explicit treatment. There are three lines of relevant evidence as follows:

1. Several individual organisms raised from birth in constant conditions of light and temperature have now been shown to manifest a spontaneous OPR: the chicken (Aschoff and Meyer-Lohmann, 1954); the lizard *Lacerta* (Hoffman, 1955); the mouse *Mus* (Aschoff, 1955). Aschoff's paper (1955) includes demonstration that the OPR is retained through three generations raised in constant conditions. Roberts (1956) has shown that a single *Drosophila pseudoobscura* raised in constant conditions of temperature and light manifests the typical OPR for that species.

2. The systems we know which are aperiodic when raised in constant conditions but can later develop an OPR, are all populations, either of individuals (*Drosophila*) or of "free" nuclei (*Neurospora*). Aperiodic cultures can be made periodic by single light stimulations which, being less than 24 hours, contain no information on the length of a solar day. There are two possible interpretations of this effect on the oscillator model: (a) we may be initiating motion in clocks initially at rest, and (b) we may be synchronizing phase in a population of clocks whose phases are randomly distributed prior to the stimulus. In a later section, detailed evidence is given to show the complete acceptability of the latter for the *Drosophila* case. But the distinction is, in any case, not crucial for the present argument. On either interpretation, we can explain the *period* of the subsequent population rhythm only in terms of an endogenous oscillation with a natural period inherent in its structure.

In her 1953 paper Harker speaks of "impressing" a rhythm on an aperiodic May-fly population by subjecting it to a "single 24-hour period." It is not clear, however, that she shows a succession of two dark-light transitions to be the minimum necessary information for producing periodicity. One of us (Pittendrigh, 1954) has shown, as Kalmus (1940b) did earlier, and Brett (1955) has recently confirmed, that a single short duration flash of light is sufficient to induce a daily OPR in previously aperiodic *Drosophila* populations. In Fig. 3 the "flash" is 4 hours long; we have other data showing a 45-second flash

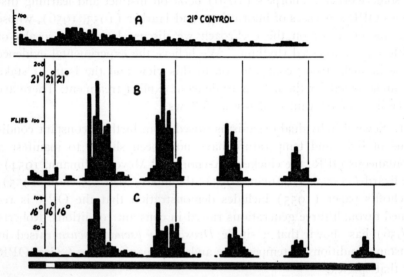

Fig. 3. "Initiation" of an eclosion rhythm in *Drosophila* by exposing a dark-grown culture to 4 hours of white light. A. Control at 21° in continuous darkness. B. 21° culture exposed to 4 hours of white light. C. 16° culture exposed to 4 hours of white light. The vertical guide lines are at $t = 0$, 4, 24, 48, 72, 96, and 120 hours after beginning the light stimulus. The ordinates represent numbers of eclosed flies per hour.

to be sufficient, and Brett uses "flashes" of one minute or more. The major point at stake is that the stimulus adequate to cause an OPR does not contain the information necessary for control of the OPR's period; and therefore the process involved cannot conceivably be called learning even in a weak sense, like that of imprinting. Learning, no matter how else it is defined, is strictly an ontogenetic accretion of information and, as such, distinct from the evolutionary accretion of information caused by selection. The clock system's information on the length of a solar day is inherited; acquisition of this information has been a phylogenetic process for which selection, not learning, has been responsible;

all that a single flash of light does is to give information on phase. Perhaps one can say the phase is "learned" but only in the crude sense that one "learns" the hour of day by inspecting, momentarily, one's watch or the position of the sun. Calhoun's (1944) statement quoted by Thorpe (1956) that we do not know of a single case of an inherited rhythm seems to us today in need of complete inversion: we do not know of a single demonstration that an organism inherits a capacity for an OPR without being able to manifest this in response to a single-step stimulus or a pulse whose duration is less than the natural period of the ESSO.

3. The third line of evidence bearing on the learning issue hinges, like the second, on the fact that an OPR does have a *natural* (innate) *period*. The observations concern the behavior of the system following the uncoupling of an external entraining periodicity. Several workers have shown that diurnal rhythms can be forced—entrained in our terminology—to a period different from 24 hours. These include *Euglena* (Bruce and Pittendrigh, 1957), *Oedogonium* (Bühnemann, 1955a, b), *Pilobolus* (Übelmesser, 1954), *Gonyaulax* (Hastings and Sweeney, 1956), *Daldinia* (Ingold and Cox, 1955), *Pseudosmittia arenia* (Remmert, 1955), and the mouse (Tribukait, 1954). Rhythms in the higher plants can also be entrained to different periods using light cycles. For references to this literature see Bünning (1956). In all but three cases discussed later, the system promptly reverts to its own natural period as soon as the external entraining agent is removed. There is no evidence, in other words, that atypical periods can be learned; just as there is none that the typical natural period is learned. The fact that periods other than the natural one can be forced on systems is fully compatible with the oscillator model; indeed, the factual detail is what one would predict on a generalized oscillator model. Here we have in mind particularly the experiments of Tribukait (1954) who shows that what we call entraining can be exercised on the mouse only within rather narrow limits on either side of the natural period.

The first of the three exceptions to our general claim is given in Übelmesser's (1954) study on the fungus *Pilobolus sphaerosporus*. She showed that the OPR (of spore discharge) can be entrained to an 8-hour period; and that following relaxation of the light-dark entraining cycle (8-hour period) the OPR maintains its precise 8-hour periodicity for exactly three cycles following which it reverts to its NP of 24 hours. This is a most complex and interesting case. We note first that the system does eventually revert sharply to its own NP just as the more typical cases do immediately. The fact that the entrained 8-hour periods last only through 3 cycles (= 24-hour total) is surely a reflection of the innate ESSO's ultimate control; but as to how this is effected we have no fully consistent interpretation.

The second exception is offered by Grabensberger's (1933-34) studies on the feeding rhythms of ants. He claims that periods other than 24 hours can be forced on the ants and persist after the training cycle is discontinued. His results are so unique and subject to the criticism of Reichle (1942-44) that we may properly await confirmation before allowing them to obscure an otherwise clear generalization.

Thirdly, there are the remarkable experiments of Pirson, Schön, and Döring (1954). There is evidently no doubt that in the alga *Hydrodictyon* a rhythm of growth and photosynthesis can be impressed with periods radically different from 24 hours (e.g. 17½) and that such periods are retained in a subsequent OPR. At present we can only say that these results are unique[4] and reflect a property of cells quite distinct from that which we designate as ESSO. It would be of first rate importance to determine the temperature relations of the *Hydrodictyon* system, and its response to single-step or pulse stimuli.

C. Entrainment phenomena. The entrainability of oscillating systems elucidates many aspects of living clocks; it also raises several warnings as to the procedure and interpretation of experiments, and it offers what seem to us to be fruitful lines of speculation.

One of the striking generalizations about persistent rhythms is that their NP, when accurately measured, is rarely an exact match for the solar day. Pittendrigh (1957) has tabulated the majority of good measurements in an earlier paper. Fig. 4 includes further examples from recent studies in our laboratory: Roberts (1956) found that in an individual of the cockroach *Byrostria fumigata* whose rhythm was precisely assayable the NP was 24 hours and 16 minutes; he finds, on the other hand, that in adults (cf. pupae, above) of *Drosophila pseudoobscura* the period is distinctly less than 24 hours (about 23 hours and 30 minutes); Burchard (1956) finds that individual hamsters have radically different periods, the majority of which are less than 24 hours

[4] A formal oscillator model could be constructed to account for the *Hydrodictyon* results. It would be based on the fact that the *period* of *free* oscillation in some nonlinear oscillators depends on the amplitude. A *forced* oscillation thus assumes an amplitude for which the period is the period of the forcing function. Free oscillations will follow removal of the forcing function and the period of the free oscillations will be the same as the previous forced period thus simulating an *apparent learning of the rhythm*.

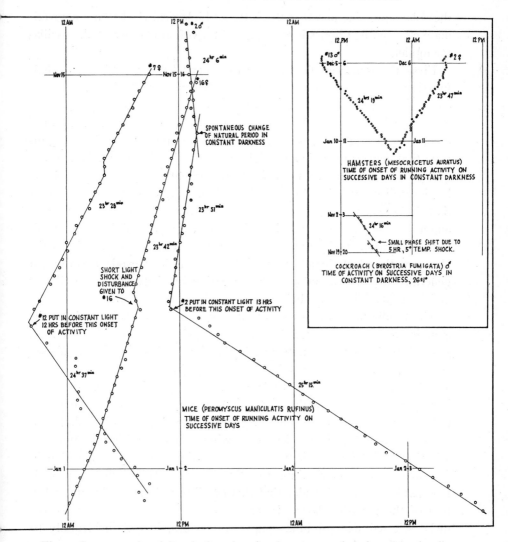

Fig. 4. Summary of activity rhythms in mice, hamsters, and cockroaches showing periods different from 24 hours. Each point on a given curve represents the time of onset of activity on a given day and the next lower point the time of onset on the following day. The slope of a line fitted through such points is a measure of the natural period. Items to be noted include the individual differences in periods, the lengthening of the period in continuous light, the existence of long periods (longer than 24 hours) in continuous darkness, and spontaneous changes in the period length.

while a few (all litter-mates so far) have NP's conspicuously longer than 24 hours (Fig. 4); Pittendrigh and Barth (1956) have similarly found occasional individuals of *Peromyscus maniculatus rufinus* with NP's more than 24 hours, although the majority are much shorter than the solar day. The fact that the NP's of biological clocks only approximate closely to 24 hours is explained in a twofold manner by the oscillator model.

First, the approximate nature of their period is compatible with their function of accurate time measurement because as oscillators they are entrainable by the light and temperature cycles of the environment. The difference between the NP of ESSO and 24 hours is the maximum error ever permitted; it is never allowed to accumulate in those organisms that sense the environmental cycle of either light or temperature. The prime importance of the light regime in entraining biological clocks has been discussed elsewhere (Pittendrigh, 1957); we need only repeat that the sole periodicity of the environment that is sharply characterized and a reliable phase giver is that of the dark-light and light-dark transitions. Entrainment by the environmental cycle of light establishes the phase of the clock in addition to controlling its period; entrainment generally involves a unique phase relationship between the oscillator and its controlling periodicity.

The oscillator model, especially its entrainment feature, explains the genetic variance encountered on the NP's of ESSO in a distinct way. It predicts, in fact, that natural selection on the NP will relax as soon as the approximation to a solar day is (a) within the entrainable range and (b) adequate for the function involved. Relaxation of selection will, however, only occur in those cases where the organism regularly sees dawn (or sunset) and is therefore subject to the control of daily environmental entrainment. In cases where this condition is not met selection will demand a better measure of the period. We note that in *Drosophila* eclosion rhythms the NP is statistically indistinguishable from 24 hours and consider that this possibly reflects the habitat selection of the pupating larva. If, as seems likely on other grounds, the niche selected will minimize pupal water loss, it will incidentally be obscured from the sun.

Other entrainment phenomena are to be anticipated on the basis of the oscillator model. Some of these bear strongly on the experimental methods involved in evaluating clock stability and precision assayed in OPR's. The likelihood that individuals in a population entrain each other in their activity cycles is the first obvious possibility; and Stephens (1957) reports that in *Uca* the OPR is indeed less stable and accurate in isolated individuals than in large groups. A similar caution on the evaluation of clock precision is raised by the phenomenon of frequency demultiplication (p. 85 above) which may take the special form of "observation entrainment" when the method of observing OPR involves energy inputs at periodic intervals that are a nearly rational fraction of the NP. Pittendrigh (1957) has already reported our encounter with this difficulty: the *Euglena* NP can in certain circumstances be stabilized by the test lights used to assay the system at 2-hour or $2\frac{1}{4}$-hour intervals; in the former case the "NP" is 24 hours, and in the latter case either $22\frac{1}{2}$ hours or $24\frac{3}{4}$ hours.

Both mutual entrainment and frequency demultiplication invite speculation on the mechanism of organic clocks. Mutual entrainment in general is a potential mechanism of stabilization and error control that may have very wide application to biological systems. Thus in our problem the stability and precision of the clock system which are two of its most challenging features could well be due to the mutual control exercised by a population of coupled ESSO's within the cell, or between cells at the tissue level of organization. The possibility that ESSO is present in replicate in the total system of the clock has another property which recommends it. Unpublished observations show that an actively dividing *Euglena* population retains clock phase over several days in spite of the fact that the generation time is shorter than the natural period of the clock. It is difficult to envisage a duplication mechanism for ESSO, no matter what its physical nature, that is instantaneous and includes registration of phase. The difficulty may not be insuperable, but at any rate the problem seems simpler if we envisage an individual cell possessing many ESSO's, each of which has had to acquire phase subsequent to its origin by duplication; and acquire it by entrainment from the rest of the cell's ESSO's to which it becomes coupled. Cell division, as in *Euglena*, can then proceed with retention of phase because the cytokinesis simply separates approximately equal halves of the population into the daughter cells.

Mutual entrainment of ESSO's with compensating temperature coefficients has been suggested (Pittendrigh, 1957) as a possible mechanism for achieving temperature independence. This feature is surely not a reflection of a single limiting physical process (diffusion or quantum effect in a large molecule) as testified by the general sensitivity to temperature perturbation (Pittendrigh, 1954) and entrainment.

Entrainment of the frequency demultiplication type could also play a useful role in the organization of the clock. One of the features which seems to present a major problem for explanation is the extremely long natural periods that have had to evolve in the ESSO's of living systems. These periods are, after all, the immense ones of stellar or planetary motions; and times of this order of magnitude seem intuitively foreign to the reaction kinetics of cellular processes. To be sure the empirical fact is that biochemical systems have evolved oscillations with these periods which we usually take as characteristic of celestial mechanics; but we retain the impression that in a chemical system variance might well increase as the natural period becomes greater. But again there are facts to be faced: Rawson (1956) shows the error in the *Peromyscus leucopus* clock to be of the order 1 in 1000. Specifically, the NP in one individual is 23 hours, 54 minutes, ± 1.6 minutes. (Note also the precision illustrated in Fig. 4.) Recognizing that the long period oscillation does exist, we still suspect that its error control, while surely endogenous to the organism, is not inherent in the long term periodicity. And we suggest that the over-all clock system may involve a hierarchy of shorter period oscillations with proportionately smaller relative variance on the period; and that these higher frequency constituents serve to stabilize, through frequency demultiplication, the long period oscillation defining the celestial period to be measured. This model incidentally presents obvious possibilities for the knotty problems involved concerning a continuously consulted clock such as those used in navigation; the fractions of the over-all celestial period may be read off from lower levels in the hierarchy of frequencies.

Finally, it is interesting to note that frequency demultiplication is a potential and testable explanation of a striking phenomenon detected first by Rawson (1956) and later independently by Burchard (1956) in the hamster clock, and Pittendrigh and Barth (1956) in the *Peromyscus* clock. The NP in these forms is very clearly defined by the time they initiate running wheel activity in constant conditions (Fig. 4). It is common to find that in an individual animal the period, which is remarkably stable over days or weeks, suddenly changes to a new value at which it again remains stable. It might be said that these changes in the NP of ESSO reflect an innate instability in keeping with its biological nature. But changes in the individual's NP have this curious feature: they consist of quite sharp transitions from one stable state to another stable state. This quality of the instability suggests to us that something akin to frequency demultiplication is responsible for the precision maintaining the stable intervals. The transition from one NP to another (Fig. 4) is thus considered to be the long period ESSO slipping in its entrainment from one whole multiple to another of the higher frequency oscillation which is stabilizing it. This interpretation is open to a quantitative test in which one examines the prediction that the observed changes of period are all whole multiples of a common time interval. The data available are still too few for the analysis demanded.

  D. *Single perturbation phenomena.* Single discrete disturbances or perturbations applied to an oscillating system will quite generally shift its phase; and as noted earlier the ultimate phase shift may or may not be preceded by a transient motion that lacks the characteristic natural period of the steady state. Earlier experiments on persistent rhythms have made comparatively little use of single discrete stimuli of the step or pulse type; the demonstration that OPR's are phase-independent of RPV's has usually, though not always, been accomplished by first subjecting the system to an inverted light cycle, as though information on period as well as phase had to be supplied exogenously. When excluding the learning interpretation we noted that single stimuli containing no information on the NP were effective in inducing OPR's in previously aperiodic systems. Figs. 5 and 6 summarize a long series of experiments. They demonstrate that the initiation of a population rhythm by a single stimulus may properly be interpreted as due to synchronization of the *aperiodic population's* clocks which are in motion but out of phase until all are simultaneously subjected to a single phase setting perturbation. These experiments form the basis of so much of our thinking about clocks that their further consideration here (cf. Pittendrigh, 1957) seems desirable.

  We consider Fig. 5 first. The solid histogram lying at 45° at the bottom of the figure shows *initiation* of a rhythm in an aperiodic culture subjected to a single perturbation using white light for 12 hours. The phase of the initiated rhythm is that normally encountered in nature or in an experimental environment of LD, 12:*12*: the peak of activity in the OPR occurs at a time corresponding to the hours immediately following the time when "dawn" would have occurred, had the system been left in LD.

  The upper part of the figure—square to the page—presents the results of what we call resetting experiments in which periodic cultures of known phase are subjected to single perturbations that *reset* the phase. The plotted points are mean times of eclosion calculated for each peak or burst of eclosion activity. There are twelve rows of plotted points, each row showing the periodicity of eclosion activity of separate cultures through a period of 96 hours; observations were made hourly. At time D_1 all twelve cultures had identical phase established by the LD cycle in which they were raised. At D_1 the light period of the LD cycle would normally have begun, but the cultures are placed in condi-

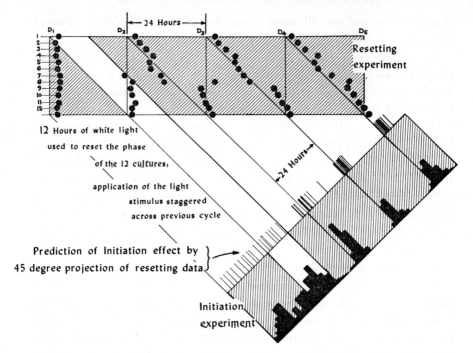

Fig. 5. The effects on the *Drosophila* eclosion rhythm of single perturbations with light (12 hours duration).

The upper figure (horizontal) shows the behavior of 12, initially synchronous, periodic cultures following perturbations that are initiated at successively later points (2-hour intervals) through the original cycle. Plotted points are means indicating the times of the periodic peaks.

The lower figure (at 45°) consists of a histogram showing the initiation by a 12-hour light stimulation of a 24-hour rhythm in a formerly aperiodic culture. Above it are lines that are 45° projections from points in the upper figure. They represent the distribution of phases in a "synthetic" population that is initially aperiodic and made periodic by a single light stimulus.

Note that in cultures number 7 and 8 the eclosion distribution between D_2 and D_3 is bimodal (two plotted points) ; in culture number 8 it is still bimodal between D_3 and D_4.

tions of darkness and temperature that remain constant except for the subsequent single 12-hour light stimulus which each receives. The stimulus is applied to the 12 cultures at successively later stages (2-hour intervals) so as to scan the whole cycle of the original rhythm. We are interested in the subsequent phase changes caused by the stimuli, as shown by the distribution of the plotted points. The vertical lines (D_2, D_3, D_4) indicate successive 24-hour intervals following D_1; the lines at 45° indicate the 24-hour intervals following initiation of the light stimulus.

By making a 45° projection of the phases of the 12 cultures we synthesize a model population that is initially aperiodic and *known* to comprise ESSO's in motion but with phases randomly distributed through the 24-hour period. The projection creates in fact the conditions we postulate exist in the aperiodic culture prior to single stimulus initiation. The 45° projection of the "resets" also renders the staggered light perturbations synchronous for the artificial population; and it shows the perturbation to be effective in resetting or synchronizing the randomly distributed phases so as to produce a clear 24-hour rhythm.

The first general point we wish to make is that the phase of the induced rhythm in the "synthetic" population duplicates precisely the phase of the rhythm in the "real" case. The strength of the whole interpretation is greatly increased by the comparable array of experiments—both initiation and resetting—in which 4-hour light perturbations are used. This whole line of work was in fact begun with the 4-hour initiation experiments which produced the remarkable result indicated in Fig. 6 and stated earlier to be unexplained (Pittendrigh, 1954, note 16). The phase of the initiated rhythm is highly atypical; eclosion peaks occur about 18 hours after the onset of light instead of about

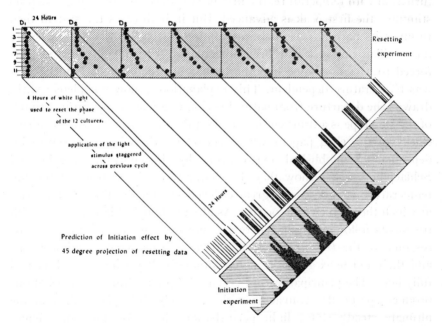

Fig. 6. The effect on the *Drosophila* eclosion rhythm of single perturbations with light (4 hours duration).

The figure is organized in the same fashion as Fig. 5; the only difference is in the duration of the light stimulus.

25 hours (or 1) as in the FR. The resetting experiments involving perturbations by 4 hours of light yield a synthetic population (when projected) that again duplicates precisely the highly characteristic phase of the 4-hour initiation experiment. It is clear that the extent of the steady-state phase shift is a function of the magnitude of the perturbation. Less extensive data on hand indicate that a duration-intensity reciprocity is involved to some extent.

Analysis of Fig. 6 yields several further points of importance. Clearly defined transients follow the perturbation before the system reaches steady state with new phase: the disappearance of the transients is reflected in a decrease of the variance on the peaks in the synthetic population. (This is an effect seen in some but not all of the initiation experiments we have done.) These transients are of interest in two distinct ways.

First, their development after disturbance of the system with single light pulses has led us to view the effects of single temperature perturbations in a very different way from that described in Pittendrigh's (1954) earlier discussion of them. Following a single step-down perturbation with temperature, the first peak is delayed; following a step-up stimulus, the first peak is advanced. But in both cases the system then resumes its natural period. In attempting to explain these facts before the oscillator model had been formulated, Pittendrigh (1954) was forced to a cumbersome hypothesis of a special "terminal clock" that was temperature dependent. This explanation is now completely withdrawn; the departure from natural period following temperature stimuli of the step type is a transient imposed on the same system (the "primary clock" of the 1954 paper) whose period is otherwise temperature independent. Unpublished experiments by Pittendrigh, Angelo and Schlaepfer (1956) show that the character of the transient following temperature step-downs is greatly dependent on the time in the cycle at which the perturbation occurs. As already noted, this is the case with transients following light. There are, however, great differences in the response of the *Drosophila* clock to light and temperature stimulation; and their existence must be noted although they cannot be discussed fully here. The principal feature is that sharp temperature perturbations produce spectacular transients with relatively little phase shift in the ultimate steady state; light perturbations have, on the other hand, spectacular effects on the steady-state phase shift.

The failure to anticipate the transients, which the oscillator model loosely predicts, misled other workers besides Pittendrigh to conclusions that are now clearly untenable. As noted previously, the widespread impression of temperature dependence in OPR's (see e.g. Bünning, 1935) was based on observation of transients, not steady state (Pittendrigh, 1954 and 1957; Leinweber, 1956).

Interest also attaches to the transient phase shifts shown in Figs. 5 and 6 for quite different reasons. Their character changes in a well-defined pattern related to the time in the old cycle at which the perturbation was initiated. Consider first Fig. 5 on the effect of a 12-hour light stimulus. Phase is reset immediately when the signal begins in the first 7 hours of the old cycle; the transient has a "period" longer than 24 hours. Signals initiated in the middle 9 hours fail to reset phase immediately; the new steady state is attained only after three transient cycles in which the "period" is more than 24 hours. Signals falling in the last 8 hours also initiate transients that endure for several cycles; but the "period" of these transients is less than 24 hours. A similar pattern is detectable following 4-hour light stimuli, but there are marked differences. The break point after which the transients are less than 24 hours occurs later in the old cycle; and the steady-state phase shift of all points is less than that developed by 12-hour perturbations.

These observations have encouraged us to believe that the descriptive oscillator model might be developed quantitatively to a point where it guides the search for a concrete model. Thus, a given ESSO must possess a characteristic wave form which, if known, should suggest its kinetic basis. It is our belief, largely intuitive at present, that features of the wave form of ESSO could be deduced from patterns like those given by Figs. 5 and 6, for the pattern relates to differences in the state of ESSO at points throughout its cycle. Present work in our laboratory is directed at further characterizing the pattern in *Drosophila* by use of different perturbation durations and intensities; and a simultaneous study is attempting to discover the comparable pattern in *Euglena* and other organisms. The basic pattern is thus regarded as a measure of ESSO itself and a tool for testing our broad hypothesis that ESSO's in diverse organisms have a fundamentally similar nature including wave form.

IV. TOWARDS A CONCRETE MODEL OF ESSO

The descriptive model of an endogenous self-sustaining oscillation as the basic time-measuring element in biological clocks serves us well in ordering the existing data and stimulating a particular line of experimental approach. But the ultimate aim, of course, is to elucidate the concrete physiological nature of ESSO. We may distinguish four broad approaches to the problem: (1) the search for potential concrete models in known oscillating systems; (2) inquiry into the level of biological organization involved; (3) extension of the formal model in a qualitative and quantitative way as a guide to the concrete mechanism; and (4) diverse direct attacks on the chemical and physical nature of ESSO. The third and fourth approaches are, for convenience, discussed together in the final section of this chapter.

A. Chemical and biological models. There are many self-sustaining oscillations in biological systems in addition to the long period ESSO we are concerned with. We know of none, however, that are not temperature dependent, like heart-beat, although this does not preclude their type of system from being a constituent in ESSO. However, for none of them is there a concrete model explaining the basic periodicity; and these offer us, therefore, little help.

Chemical periodicities are also fairly common (Hedges and Myers, 1926). The best known of these is the phenomenon of Liesegang rings. It, too, is of limited interest for us being a periodicity in space and dependent on differential diffusion rates. Undamped periodic chemical reactions are known and have been subjected to a theoretical kinetic description by Lotka (1920). To the best of our knowledge, however, it has never been shown that any real chemical periodicity does take the kinetic form given by Lotka's equations.

Galston and Dahlberg (1954) have suggested that the mechanism underlying endogenous daily rhythms in plants is a periodicity of auxin caused by the adaptive enzyme IAA-oxidase. This first struck us as an ingenious application of the Lotka-Volterra type of prey-predator kinetics to the molecular level; but Lotka evidently developed his equations, best known to the biologist in their applications to prey-predator oscillations, for the chemical case. The Galston-Dahlberg model involves assumptions on the kinetics of formation and destruction of both IAA and the oxidase that destroys it, which are probably not justified. Furthermore, the subsequent work of Ball and Dyke (1956) as well as the recent realization that the system to be explained is temperature independent (Bünning and Leinweber, 1956) render the model, apart from its assumptions, inadequate. A non-specific formal chemical model for endogenous rhythms was proposed by Mori (1947). Mori's model involves a series of reactions both reversible and irreversible, and he compares his hypothesis to complicated electronic oscillators.

There is considerable literature on electrochemical oscillations, the prime example of which is the iron-wire model of nerve activity. Franck and Meunier (1953) and Franck (1956) have discussed several of them. They have, in particular, shown how these electrochemical oscillations may be coupled in various ways and exhibit mutual entrainment that has a remarkable parallel in some biological oscillations. Franck's paper (1956) develops a line of thought similar to our own; he introduces the concept of a population of close-lying oscillating electrodes (which could be synchronized under certain conditions) as a model for electroencephalograms observed during certain pathological states. Entrainment phenomena have also been invoked by Pringle (1952) in a stimulating discussion that seeks a model of learning based on a population of oscillations.

In spite of its great intrinsic interest, this literature seems to offer us little immediate help especially in relation to the major problems— temperature independence; and the length, stability, and precision of the natural period.

B. Level of biological organizations. The higher metazoa, especially arthropods and vertebrates, have attracted much attention in the study of persistent rhythms. All the spectacular cases of continuously consulted clocks are in vertebrates or arthropods; and these two groups have also yielded the bulk of those rhythms remarkable for their stability and precision. It is not surprising that the nervous system has been a favorite candidate for the physical basis of clocks in these forms. There is an a priori attractiveness to a system that is known to manifest entrainable spontaneous rhythmic discharges and that can include reverberatory circuits and summating devices. Little difficulty should be anticipated in conceiving a clock-like neurological mechanism that exploits all the features of mutual entrainment and frequency demultiplication referred to, speculatively, above. And such a model could, of course, incorporate the feature of temperature independence through mutual entrainment of component neural oscillations with complementary temperature coefficients.[5] There is, of course, no question that the nervous system forms part of the input and output "black boxes" of the total system underlying metazoan OPR's. And Harker's (1956) recent and elegant experiments with the cockroach *Periplaneta americana* have demonstrated that the clock itself (ESSO) is physically located in the

---

[5] A recent paper by Kerkut and Taylor (1956) describes interesting transients in spontaneous discharge frequencies following step-up and step-down temperature stimulation. Especially interesting is the fact that the transients show negative temperature coefficients.

nervous tissue of this species. Harker has demonstrated that headless (but living) roaches are arhythmic but reacquire a rhythm following implantation of a sub-oesophageal ganglion from a rhythmic donor roach. The crucial feature in Harker's demonstration is that the ganglion brings with it the phase of the donor's rhythm.[6] This is one of the most remarkable experiments in the field of persistent daily rhythms.

Harker's experiment should, however, not be misinterpreted as evidence for the necessity of the nervous system. Several recent papers support our assumption that the endogenous daily rhythms in plants are true biological clocks based on an ESSO like those in animals. Ball and Dyke (1954) have demonstrated temperature independence in the growth rhythm of the *Avena* coleoptile. And the extensively studied rhythm of leaf movement in *Phaseolus*, formerly believed to be dependent on temperature, has also been shown now to have a steady-state period that is independent of the temperature (Bünning and Leinweber, 1956; and Leinweber, 1956). Thus the ESSO for which we seek a concrete model exists in organisms without a nervous system.

It is, furthermore, known now that a single cell can manifest an OPR with all the properties familiar in the higher plants and animals. In particular, temperature independence has recently been demonstrated in an OPR in *Euglena* (Bruce and Pittendrigh, 1956) as well as in *Gonyaulax* (Hastings and Sweeney, 1956) and *Paramecium bursaria* (Ehret, 1956). The cellular level of organization is, therefore, sufficient for a clock. This conclusion seems to us of first rate importance. The basic mechanism of ESSO is to be sought in the physical chemistry of subcellular parts and processes, and not, for instance, in the attractive possibilities offered by a population of neurones.

We do not mean to imply here that a single cell constitutes the clock in higher metazoa; nor that their nervous system is not involved. It is our view that probably all cells are autonomously capable of an ESSO with the long period of a solar day; but that in higher animals the function of daily time measurement is executed by a few that are specialized, in typical metazoan fashion, for this particular function.

[6] Harker's paper states that implantation of the ganglia carries in ". . . the rhythm of activity previously shown by the donor of the sub-oesophageal ganglia." We take this wording to mean that the ganglion brings with it not only the period but also the phase of the donor roach's rhythm. Transmission of phase is crucial in an experiment of this kind; until it has been demonstrated, a less interesting conclusion is not excluded. This alternative is that the initial decapitation of the roach destroyed essential endocrine output black boxes (cf. Fig. 1) which are restored when a sub-oesophageal ganglion is returned by implantation.

And we believe further that the metazoa have surely exploited the evolutionary opportunity for elaboration and stabilization of time measurement at the higher level of organization afforded by their many cells. Thus we have no doubt that the elaborate chronometers of birds, bees, and crustaceans are all in the nervous system and probably depend on neurological organization for some of their unique complexity. But we emphasize our view that the basic mechanism has only been elaborated, not created, in the course of metazoan evolution—that it is present, posing the fundamental problem, in the organization of a single cell.

C. Difficulties confronting direct approaches. It is important to distinguish the legitimate conclusion that a single cell *is* a clock from the alternative wording that it *has* a clock. Thus it remains to be shown that the cell *has* a clock which is a physically distinct or isolatable component as against a feature of its total organization. For reasons largely intuitive, we are inclined to the latter view. But it is clearly desirable to extend delimitation of the necessary and sufficient level of complexity as low as possible; the lowest level necessary for ESSO is the upper level of complexity allowable to the model. Further delimitation might be attempted in various ways; an obvious first question is "nucleus or cytoplasm?" Dr. Bünning is evidently thinking along these lines, as his paper in this volume shows.

The question "nucleus or cytoplasm?" raises a serious difficulty to be faced in all such attacks on the concrete nature of ESSO. Safe conclusions can be drawn from conceivable outcomes of experiments in which nuclei and cytoplasm are brought together after prior exposure to conditions that establish a different phase in each. When this is done, as we are preparing to do in *Neurospora* heterokaryons, the *autonomy* and hence sufficiency of either constituent can be evaluated from its ability to control the phase of the compound. In short, we need techniques analogous to Harker's in which we demonstrate clock autonomy in a manipulatable part.

There are other possible observations, attractive for their simplicity, whose inadequacy needs emphasis. For instance, to justify the conclusion that the clock is nuclear, it is insufficient to demonstrate a periodicity in some nuclear feature (like volume). Our comment on so specific a point is justified by the very general issue involved: the principal problem in observational clues to the concrete nature of ESSO is that of distinguishing a periodicity which is ESSO itself (or part of it) from a periodicity that is caused by ESSO. It is this difficulty which discourages a blunt attack on the problem by, for instance, chromatographing cells like *Euglena* systematically through the 24-hour period. Periodicity of chemical features of the cell is a priori certain and empirically known; what is not known is the meaning of the observed cycles and what is needed are criteria for their interpretation. Thus, it seems to us that direct chemical observation should be deferred until it takes the form of testing strong hypotheses inspired by other approaches.

We have already suggested that quantitative development of the formal oscillator model might lead to such an hypothesis, and it remains now to indicate why patterns of transients and phase shifts appeal to us as the most promising way to develop the model. Interpretation of all other features of the OPR as a reflection of ESSO kinetics is beset with difficulties arising from two considerations: (1) There is never any certainty even for a single cell that the clock comprises only one ESSO; and (2) the physiological "black boxes" on both the input and output sides of ESSO are of unknown nature in most cases; and in all cases, especially on the output to OPR, they undoubtedly transform information in a manner obscure to the experimenter.

The first difficulty is especially relevant to interpreting the kinetics of initiation, amplitude, damping, and loss of the OPR; for all these features of OPR are open to explanation in terms of the gain and loss of synchrony in a population of ESSO's—even in the single cell (cf. p. 87 above).

The second difficulty is just as severe and in some respects more so. At first sight the study of factors (like drugs, light, etc.) effecting a change in the NP seems a promising tool, for the NP of the total system is, on hypotheses, the one feature inherent in ESSO's structure and independent of the input complications. But the matter is not so simple; conclusions cannot be drawn from the period changes wrought by light or drugs until we know that the *primary* action of these factors is on ESSO itself and not mediated by unknown reactions whose products interfere with ESSO kinetics. In the case of light, which does produce spectacular effects on the steady-state period in vertebrates (cf. Fig. 4), we can be sure the primary action is not on ESSO; blinded mice are normally rhythmic (the effective ESSO is not in the retina), but their phase and period are insensitive to the light regime (Whitaker, 1940). Even in a single cell the likelihood of unknown input boxes between the receptor pigment and ESSO remains to discourage undertaking the labor of an action spectrum; for the spectrum obtained clearly cannot be taken as characteristic of a component in ESSO itself.

In conclusion we return to the pattern of phase shifts and transients caused by single signals as a preferred tool to investigate the wave form of ESSO on which they presumably depend. This method cannot be proven free from the objections raised to others; it is only our feeling that this is likely to be true. The extent to which ESSO inputs and outputs determine the pattern can be estimated to some extent in our comparative approach; for there is slight chance that the "black boxes" are identical in the control of activities as diverse as *Euglena* phototaxis, *Neurospora* conidiation, mouse running activity, and *Drosophila* eclosion. The comparative approach in general recommends itself as the most hopeful method of elucidating the properties of the ESSO we suppose is common to cells.

V. SUMMARY

The outlook developed in the present paper can be summarized as follows:

1. The widespread phenomena of temperature-independent persistent daily rhythms is regarded as a manifestation of the same basic time-measuring ability that underlies the elaborate chronometry of birds and other metazoa; it is maintained, in fact, that temperature-independent time measurement is a nearly universal feature of living systems.

2. A descriptive oscillator model and terminology is developed and shown to be adequate for interpretation of all known effects and fruitful of speculative leads; the basic time-measuring element in all biological clocks is described as an *endogenous self-sustaining oscillation* (ESSO), probably of the relaxation type.

3. It is shown that the search for a physiological explanation of ESSO must by-pass the learning phenomenon and concern itself with the inherited, or innate, features of the cell's physical and chemical organization; the cellular level (vis-à-vis tissue or organ level) of complexity is found to be sufficient to develop a clock.

4. The search for a concrete model of ESSO is confronted with the following difficulties:

(a) Uncertainty as to whether ESSO is a physically distinct part of the cell or, on the other hand, a feature of the over-all organization of the cell; and, consequently,

(b) the operational problem of distinguishing a periodicity that *is* ESSO (or an ESSO component) from other cellular (or organismic in the case of metazoa, etc.) periodicities that *are caused by* ESSO.

It is argued that a comparative study of persistent rhythms in several very diverse organisms is a promising tool for making this crucial distinction.

BIBLIOGRAPHY

Aschoff, J. 1955. Tagesperiodik bei Mäusestämmen unter konstanten Umgebungsbedingungen. *Pflüger's Archiv. 262*, 51-59.

Aschoff, J., and J. Meyer-Lohmann. 1954. Angeborene 24-Stunden Periodik beim Kücken. *Pflüger's Archiv. 260*, 170-176.

Ball, N. G., and I. J. Dyke. 1954. An endogenous 24-hour rhythm in the growth rate of the *Avena* coleoptile. *J. Exp. Bot. 5*, 421-433.

Ball, N. G., and I. J. Dyke. 1956. The effects of indole-3-acetic acid and 2:4 dichlorophenoxyacetic acid on the growth rate and endogenous rhythm of intact *Avena* coleoptiles. *J. Exp. Bot. 7*, 25-41.

Bateman, M. A. 1955. The effect of light and temperature on the rhythm of pupal ecdysis in the Queensland fruit-fly, *Dacus (Strumeta) tryoni* (Frogg.) *Austral. Jour. of Zool. 3*, 22-33.

Bethe, A. 1940-41. Die biologischen Rhythmus-Phäenomene als selbständige bzw. erzwungene Kippvorgänge betrachtet. *Pflüger's Archiv. 244*, 1-43.

Brett, W. J. 1955. Persistent *diurnal* rhythmicity in *Drosophila* emergence. *Ann. Entom. Soc. Amer. 48*, 119-131.

Brown, F. A., Jr., and H. M. Webb. 1948. The temperature relations of an endogenous daily rhythmicity in the fiddler crab, *Uca. Physiol. Zool. 21*, 371-381.

Brown, F. A., Jr., J. Shriner, and C. L. Ralph. 1956. Solar and lunar rhythmicity in the rat in 'Constant Conditions' and the mechanism of physiological time measurement. *Am. J. Physiol. 184*, 491-496.

Brown, F. A., Jr., H. M. Webb, and M. F. Bennett. 1955. Proof for an endogenous component in persistent solar and lunar rhythmicity in organisms. *Proc. Nat. Acad. Sci. 41*, 93-100.

Bruce, V. G., and C. S. Pittendrigh. 1956. Temperature independence in a unicellular "clock." *Proc. Nat. Acad. Sci. 42*, 676-682.

Bruce, V. G., and C. S. Pittendrigh. 1957. Endogenous Rhythms in Insects and Microorganisms. Symposium on Biological Chronometry at the A.I. of B.S. Symposium in August, 1956. *Amer. Naturalist*, suppl. (in press).

Bühnemann, F. 1955a. Die rhythmische Sporebildung von *Oedogonium cardiacum* Wittr. *Biol. Zbl. 74*, 1-56.

Bühnemann, F. 1955b. Das endodiurnale System der *Oedogonium*-Zelle. III. Über den Temperatureinfluss. *Zeit. Naturforschg. 10b*, 305-310.

Bünning, E. 1935. Zur Kenntnis der endonomen Tagesrhythmik bei Insekten und bei Pflanzen. *Berichte d. Deutschen Bot. Ges. 53*, 594-623.

Bünning, E. 1956. Endogenous rhythms in plants. *Ann. Rev. Plant Physiol. 8*, 71-90.

Bünning, E., and F. J. Leinweber. 1956. Die Korrektion des Temperatur-fehlers der endogenen Tagesrhythmik. *Naturwiss. 43*, 42-43.

Burchard, J. E. 1956. Unpublished experiments done in this laboratory.

Calhoun, J. B. 1944. Twenty-four hour periodicities in the animal kingdom. *J. Tenn. Acad. Sci. 19*, 179-200, 252-262.

Ehret, C. L. 1956. Personal communication.

Franck, U. F. 1956. Models for biological excitation processes. *Progress in Biophysics and Biophysical Chemistry 6*, 171-206.

Franck, U. F., and L. Meunier. 1953. Gekoppelte periodische Elektroden-vorgänge. *Zeit. f. Naturforsch. 8*, 396-406.

von Frisch, K. 1950. Die Sonne als Kompass im Leben der Bienen. *Experientia 6*, 210-221.

Galston, A. W., and L. Y. Dahlberg. 1954. The adaptive formation and physiological significance of indole-acetic-acid-oxidase. *Am. J. Bot. 41*, 373-380.

Grabensberger, W. 1933-34. Untersuchungen über das Zeitgedächtnis der Ameisen und Termiten. *Zeit. f. vergleich. Physiol. 20*, 1-54.

Harker, J. 1953. The diurnal rhythm of activity of mayfly nymphs. *J. Exper. Biol. 30*, 525-533.

Harker, J. 1956. Factors controlling the diurnal rhythm of activity of *Periplaneta americana. J. Exper. Biol. 33*, 224-234.

Hastings, J. W., and B. M. Sweeney. 1956. Personal communication.

Hedges, E. S., and J. E. Myers. 1926. *The Problem of Physico-Chemical Periodicity.* Longmans, Green and Co., New York.

Hill, A. V. 1933. Wave transmission as the basis of nerve activity. *Cold Spring Harbor Symp. Quant. Biol. 1*, 146-151.

Hoffmann, K. 1954. Versuche zu der im Richtungsfinden der Vögel ent-haltenen Zeitschätzung. *Zeit. f. Tierpsychol. 11*, 453-475.

Hoffmann, K. 1955. Aktivitätsregistrierungen bei frisch geschlüpften Eidechsen. *Z. vergl. Physiol. 37*, 253-262.

Ingold, C. T., and V. J. Cox. 1955. Periodicity of spore discharge in *Daldinia. Ann. of Bot. 19*, 201-209.

Kalmus, H. 1933. Über die Natur des Zeitgedächtnisses der Bienen. *Zeit. f. vergl. Physiol. 20*, 405-419.

Kalmus, H. 1935. Periodizität und Autochronie als zeitregelnde Eigen-schaften der Organismen. *Biol. Generalis 11*, 93-114.

Kalmus, H. 1937-38. Tagesperiodisch verlaufende Vorgänge an der Stabheuschrecke (*Dixippus morosus*) und ihre experimentelle Beein-flussung. *Zeit. f. vergl. Physiol. 25*, 494-508.

Kalmus, H. 1938. Über das Problem der sogenannten exogenen und endogenen, sowie der erblichen Rhythmik und über organische Perio-dizität Überhaupt. *Rivista di Biol. 24*, 191-225.

Kalmus, H. 1940a. Diurnal rhythms in axolotl larvae and in *Drosophila. Nature 145*, 72.

Kalmus, H. 1940b. New research in the diurnal periodicity of animals. *Acta Med. Scand.*, suppl., *108*, 227-233.

Kerkut, G. A., and B. J. R. Taylor. 1956. Effect of temperature on the spontaneous activity from the isolated ganglia of the slug, cockroach, and crayfish. *Nature 178*, 426.

Kramer, G. 1952. Experiments on bird orientation. *Ibis 94*, 265-285.

Leinweber, F. J. 1956. Über die Temperaturabhängigkeit der Periodenlange bei der endogenen Tagesrhythmik von *Phaseolus*. *Zeit. f. Bot. 44*, 337-364.

Lotka, A. J. 1920. Undamped oscillations derived from the law of mass action. *J. Am. Chem. Soc. 42*, 1595-1599.

Matthews, G. V. T. 1955. *Bird Navigation*. Cambridge University Press, Cambridge, England.

Mori, S. 1947. A concept on mechanisms of the endogenous daily rhythmic activity. *Memoirs Coll. Sci., U. of Kyoto, Series B, 19*, 1-4.

Pardi, L. 1954. Über die Orientierung von *Tylos latreillii* Aud. and Sav. (Isopoda terrestria). *Zeit. f. Tierpsychologie 11*, 175-181.

Pardi, L., and M. Grassi. 1955. Experimental modification of direction finding in *Talitrus saltator* (Montagu) and *Talorchestia deshayesi* (Aud.) (Crustacea-Amphipoda). *Experientia 11*, 202-210.

Pardi, L., and F. Papi. 1953. Ricerche sull'orientamenta di *Talitrus saltator* (Montagu) (Crustacea-Amphipoda). *Zeit. vergl. Physiol. 35*, 459-489 and 490-518.

Pirson, A., W. J. Schön and H. Döring. 1954. Wachstum und Stoffwechselperiodik bei *Hydrodictyon*. *Zeit. f. Naturforschg. 9b*, 350-353.

Pittendrigh, C. S. 1954. On temperature independence in the clock-system controlling emergence time in *Drosophila*. *Proc. Nat. Acad. Sci. 40*, 1018-1029.

Pittendrigh, C. S. 1957. Perspectives in the study of biological clocks in *Symposium on Perspectives in Marine Biology*. University of California Press, Berkeley. In press.

Pittendrigh, C. S., J. Angelo and W. Schlaepfer. 1956. Unpublished experiments done in this laboratory.

Pittendrigh, C. S., V. G. Bruce, N. Rosensweig, and M. Rubin. 1957. Unpublished experiments done in this laboratory.

Pohl, R. 1948. Tagesrhythmus im phototaktischem Verhalten der *Euglena gracilis*. *Zeit. f. Naturforschg. 3b*, 367-378.

van der Pol, B., and J. van der Mark. 1929. The heartbeat considered as a relaxation oscillation and an electrical model of the heart. *Arch. Neerl. Physiol. 14*, 418-443.

van der Pol, B. 1940. Biological rhythms considered as relaxation oscillations. *Acta Med.-Scand.*, suppl., *108*, 76-88.

Pringle, J. W. S. 1952. On the parallel between learning and evolution. *Behaviour 3*, 174-215.

Rao, K. P. 1954. Tidal rhythmicity of rate of water propulsion in *Mytilus* and its modifiability by transplantation. *Biol. Bull. 106*, 353-359.

Rawson, K. S. 1956. Homing behaviour and endogenous activity rhythms. Ph.D. Thesis, Harvard University.

Reichle, F. 1942-44. Untersuchungen über Frequenzrhythmen bei Ameisen. *Zeit. vergl. Physiol. 30*, 227-251.

Remmert, H. 1955. Untersuchungen über das tageszeitlich gebundene Schlüpfen von *Pseudosmittia arenia* (Dipt. Chironomidae). Zeit. vergl. *Physiol. 37*, 338-354.

Roberts, S. K. 1956. Unpublished experiments done in this laboratory.

Semon, R. 1912. *Das Problem der Vererbung "Erworbener Eigenschaften."* Engelmann, Leipzig.

Simpson, G. G., C. S. Pittendrigh, and L. H. Tiffany. 1957. *Life.* Harcourt, Brace and Co.

Stein-Beling. 1935. About time-memory in animals. The seat of time-memory in bees. *Biol. Rev. Cambr. Phil. Soc. 10*, 18-41.

Stephens, G. C. 1957. Twenty-four hour rhythmicity in marine organisms. Symposium on Biological Chronometry at the A.I. of B.S. Symposium in August, 1956. *Am. Naturalist*, suppl. (in press).

Thorpe, W. H. 1956. *Learning and Instinct in Animals.* Methuen, London.

Tinbergen, N. 1953. *Social Behaviour in Animals.* John Wiley and Sons, New York.

Tribukait, B. 1954. Aktivitätsperiodik der Maus im Künstlichen verkürzten Tag. *Naturwiss. 41*, 92-93.

Übelmesser, E. R. 1954. Über den endonomen Rhythmus der Sporangien-träger bildung von *Pilobolus. Arch. f. Mikrobiol. 20*, 1-33.

Wahl, O. 1932. Neue Untersuchungen über das Zeitgedächtnis der Bienen. *Zeit. vergl. Physiol. 16*, 529-589.

Welsh, J. H. 1938. Diurnal rhythms. *Quart. Rev. Biol. 13*, 123-139.

Whitaker, W. L. 1940. Some effects of artificial illumination on reproduction in the white-footed mouse, *Peromyscus leucopus noveboracensis. J. Exp. Zool. 83*, 33-60.

VI. ENDOGENOUS DIURNAL CYCLES OF ACTIVITY IN PLANTS

BY E. BÜNNING[1]

I. INTRODUCTION

IN PLANTS many processes may be observed which show diurnal fluctuations even while the external conditions remain constant. Such an endogenous periodicity may be observed in algae and fungi as well as in higher plants. This endogenous diurnal periodicity may have the consequence of activity cycles in growth, turgor pressure (Enderle, 1951), respiration, spore discharge (Schmidle, 1951; Bühnemann, 1955a–c), and many other processes. (For a survey of these phenomena see Bünning, 1956.)

Obviously all these diurnal rhythms in the activity of physiological processes are consequences of some elementary diurnal process within the cell. At present this elementary process is unknown. Therefore, we must restrict ourselves to the study of one of the other processes being directed by this unknown one. Several examples of physiological processes which in many plants depend, to a high degree, upon a cellular endodiurnal system are: the diurnal movements of leaves and the diurnal discharge of spores in several algae and fungi. It is well known that processes like these may also be regulated by external factors. Thus constant external conditions are always needed in order to find the endogenous rhythmicity involved in these reactions.

Investigations on endogenous periodicity in plants date back some 200 years, for in 1759 Zinn published a paper on diurnal leaf movements and was able to show that these movements may persist without cycles of light and dark or of high and low temperatures. It was not before the experiments of Pfeffer (1907 and later) on the occurrence of endogenous periodicity that this phenomenon was clearly demonstrated. Much of the following discussion is based on studies we have made in *Phaseolus*. The blade of a bean leaf moves upward and downward in a diurnal rhythm, by virtue of changes in turgor of the cells of the pulvinus between petiole and blade of the leaf. We have recorded these movements by connecting the blade of a leaf to the pen of a simple kymograph. Figs. 1, 2, 3, and 8 show such records.

[1] Department of Botany, University of Tübingen, Germany.

II. LENGTH OF PERIODS

We cannot expect the endogenous periodicity to work as accurately as a watch. Quite often distinct deviations from the exact 24-hour period can be observed. Studying periodic spore discharge in *Oedogonium*, Bühnemann recorded cycles of 22 hours. In *Phaseolus* we found strains which, under standard conditions, show considerable variations in the length of the periods. These differences proved to be hereditary. At present we use a strain with periods of 28 hours (Fig. 1A). As soon as the external factors change to a diurnal pattern they exercise a regulating influence, and the leaf movements show periods of exactly 24 hours. A regulation however is possible only within certain limits (Fig. 1B). If the external rhythm has periods shorter than approximately 18 hours, periods of, for example, 22, 24, or 28 hours result due to the internal cyclical processes (Fig. 1C).

III. EVOCATION OF THE RHYTHM BY EXTERNAL STIMULI

An endogenous rhythm is a rhythm which continues while the external factors remain constant. But in order to evoke the periodicity, some external impulse is often necessary. This may be compared with the impulses necessary for the movement of a pendulum.

An impulse is, for example, required in order to start the diurnal movements in etiolated *Phaseolus*. If the plants have grown up in permanent darkness and constant temperature the leaves do not show any movements at all. A single stimulus will, however, start the diurnal movements. This stimulus may consist of a short light period of a few hours (Fig. 2). The transition of the plant from continuous darkness to continuous light or from continuous light to continuous darkness is also a sufficient impulse.

How can we explain this evocation of a periodic process by a single stimulus? We may suppose that even prior to this evocation diurnal processes are going on within the plant. But these processes are not synchronous within the different cells. In this case, the external stimulus would not have any other function than to synchronize the processes within the several cells. There is evidence in favor of this explanation. The fact that we can disturb the synchronous running of the several "clocks" is especially important. Light stimuli can suffice to desynchronize the process. The periodicity is now subdivided into several processes, each of which has periods of about 24 hours. A single strong light stimulus offered later on may cause a new synchronization (Fig. 3).

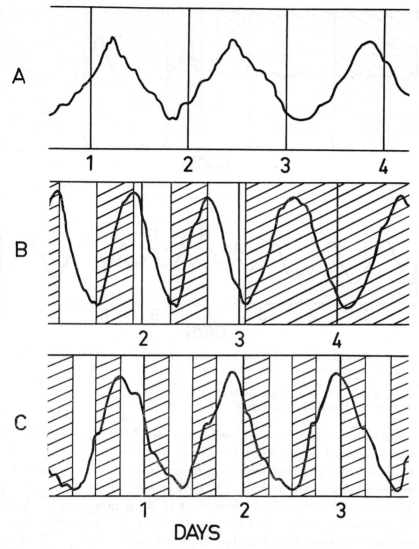

Fig. 1. Kymograph records of leaf movements of *Phaseolus multiflorus*. Abscissae: Days after beginning of recording the movements. Dark periods indicated by shading. A. In continuous light. Periods longer than 24 hours. B. Light-dark cycles of 9:9 hours regulate length of periods to 18 hours. During the following time of continuous darkness the length of periods is again the normal one of about 28 hours. C. In 6:6 hours light-dark cycles. No regulation but only the endogenous rhythm of about 28 hours.

The rules for the regulation of the internal rhythm by external factors, especially by light and darkness, have been elaborated by several research workers (Flügel, 1949). In many cases the endogenous rhythm

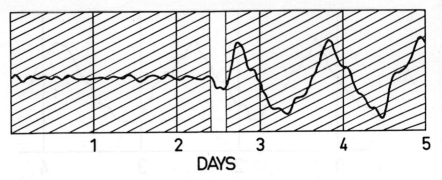

DAYS

Fig. 2. *Phaseolus multiflorus*, grown from germination in continuous darkness. Evocation of movements by a single light period. Kymograph records.

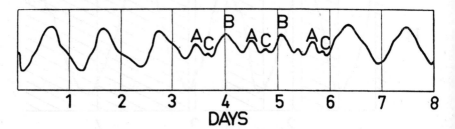

DAYS

Fig. 3. Kymograph records of leaf movements of *Phaseolus multiflorus* in continuous darkness. Desynchronization by a light stimulus given on the first day. The points AAA, BB, and CCC mark maxima of 3 partial rhythms, each of them having periods of about 24 hours. A second light stimulus at the end of the 6th day causes new synchronization. From Bünning, 1935.

reaches an extreme value approximately 16–18 hours after one of the aforementioned impulses. In *Phaseolus*, for example, this means that 17 hours after the beginning of a light period the leaves reach their maximum downward position. The next maxima are 17 + 24, 17 + 48 hours, etc., after the initial stimulus. It is rather unimportant whether this stimulating light period lasts for less than one hour or whether it continues for more than one day. With coleoptiles of *Avena*, Ball and Dyke (1954) found a maximum rate of growth at 16, 16 + 24, and 16 + 48 hours after replacing continuous red light by continuous darkness.

IV. THE ACTION SPECTRUM OF LIGHT REGULATION

A very simple method has been used to find the spectral regions which are active in regulating the endogenous rhythm. As already mentioned, a diurnal cycle of light and darkness may regulate the periods of the internal rhythm to exactly 24 hours, whereas under constant conditions the periods vary with the species between 22 and 28 hours. Now we only have to test which portions of the spectrum have this regulating effect. This experiment becomes even clearer if the plants are given periods deviating more than the 24-hour cycles from those observed under constant conditions. The previously mentioned strain of *Phaseolus* was used which, under constant conditions, has periods of 28 hours. These plants received cycles of 10 hours light and 12 hours darkness. That means we tried to force the plant to a 22-hour rhythm.

The result was quite astonishing (unpublished data). Most parts of the visible spectrum are not effective. There is some effect of blue light, no effect of ultraviolet. A strong effect is, however, produced by the near infrared with wave lengths beginning at about 700 mμ. Wave lengths of more than 775 mμ are no longer effective. Although we could not yet measure the action spectrum in detail, apparently the same pigment system is involved which is also responsible for photoperiodic reactions and several other light reactions in plants. This pigment system shows, as it is well known, a maximum at 735 mμ. This result seems to be quite interesting as there exists a close relation between endogenous periodicity and photoperiodism.

Experiments of Bühnemann (1955c) on the spore discharge in the alga *Oedogonium* also revealed the regulating effect of red or near infrared (though this author draws another conclusion from his results). Ball and Dyke (1954) in experiments on growth rhythms in *Avena* coleoptiles found that the transition from continuous infrared to darkness was effective in starting the periodicity.

V. INFLUENCE OF TEMPERATURE ON THE LENGTH OF PERIODS

In order to understand the mechanism of the endogenous diurnal periodicity we tried to disturb this clock system by several factors. When considering the influence of temperature one would expect a more rapid endogenous rhythm at higher temperatures and the contrary at lower temperatures. Experiments with *Phaseolus* published about 25 years ago are in agreement with these expectations (Bünning, 1931). The results of these early experiments are shown in Table I below. From these results a Q_{10} of about 1.2 was calculated. More recently we obtained quite different results by allowing the plants also to grow up at different temperatures (Bünning and Leinweber, 1956; Leinweber, 1956). Now the plants were kept, beginning with seed germination, at the respective temperatures. The results are summarized in Table II.

TABLE I. *Phaseolus multiflorus.* Prior to the recording of the leaf movements at different temperatures the plants were kept at 20°C.

Temperature	Length of Periods hours
15°C	29.7
20°C	27.0
25°C	23.7
30°C	22.0
35°C	19.0

TABLE II. As Table I, but plants were grown from the time of germination at experimental temperatures.

Temperature	Length of Periods hours
15°C	28.3 ± 0.4
20°C	28.0 ± 0.4
25°C	28.0 ± 1.0

Under these conditions the length of the periods is nearly independent of the temperature. Therefore the plant is able to compensate the influence of temperature if it is accustomed to it. To determine how quickly this compensation is working, plants were grown at normal medium temperature, and it was not before beginning to record the leaf movements that they were exposed to the extreme high or low temperatures. Observation was made of the number of days which elapsed until the influence of temperature was compensated. If, for example, the plants are brought to low temperatures they show at first a retardation of the process which results in very long periods. After a few days the adaptation already becomes visible, and the periods become shorter. For a certain time they are even shorter than normal. This is illustrated by the results summarized in Table III.

TABLE III. *Phaseolus multiflorus.* Length of periods (hours) after transfering the plants from the greenhouse to 15° or 25°C.

Temperature	1.	2.	3.	4. day
15°C	33.4 ± 0.4	26.2 ± 0.6	24.6 ± 0.9	24.1 ± 1.3
25°C	28.8 ± 0.5	25.1 ± 0.6	26.8 ± 0.6	26.6 ± 1.0

Apparently after exposing the plants to a low temperature some process of adaptation starts.

The same results are obtained if the plants are exposed for only a short time to low temperatures and immediately after to the temperature which prevailed previously. Such an experiment is shown in Table IV.

TABLE IV. *Phaseolus multiflorus*. Leaf movements recorded at 15°C, but temperature brought to 1°C for several hours. Length of periods (hours).

Day before 1° -treatment	Day of 1° -treatment	1 Day Later	2 Days Later	3 Days Later
28.4 ± 0.8	32.1 ± 0.8	24.4 ± 0.8	24.7 ± 0.8	20.9 ± 0.9

Again the retarding effect of low temperature and the induction of a process of compensation which causes an acceleration is observed.

According to these observations the mechanism of the endogenous periodicity involves several processes with different temperature coefficients. The whole process is made independent of temperature by a variation in the quantitative relation between the several processes. The process of adaptation is of great importance in this correlation. If the correlation works inaccurately a clear temperature dependency is involved. This may entail either acceleration or retardation by increased temperature. A retardation was found by Bühnemann (1955b) for *Oedogonium*. He found the following length of periods:

$$27.5°C: 25 \text{ hours}$$
$$17.5°C: 20 \text{ hours}$$

Thus Q_{10} in this case is 0.8.

There are facts indicating that the plant has difficulties in maintaining the exact functioning of its clock at extreme temperatures. The mechanism fails very often if the temperature deviates too much from the normal (Bünning, 1932; Leinweber, 1956). This failure may be recognized by measuring the number of cases in which the periodicity fails, after the plant has been returned to constant conditions (Fig. 4). If during these constant conditions the temperature is higher than optimum we may recognize the rhythmicity for one week or longer, but at extreme temperature it fades very quickly.

VI. RELATION TO METABOLISM

It is evident that processes like leaf movement or spore discharge, which can be recorded without difficulties, are not identical with the very nature of the endogenous periodicity. They are just consequences of this periodicity and, moreover, depend also on other factors. The

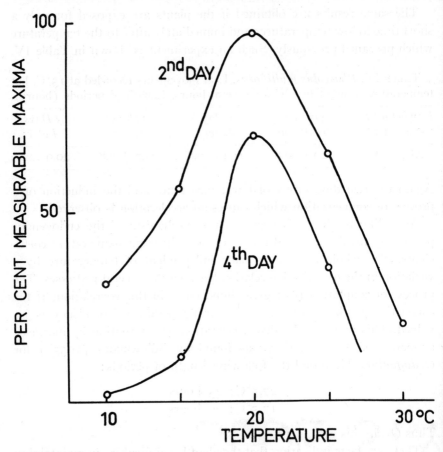

Fig. 4. Leaf movements of *Phaseolus multiflorus* grown in the greenhouse. The curves illustrate the percentage of plants with distinct diurnal movements at the 2nd and 4th day after removal to constant conditions. The periodicity is best maintained if the temperature during these constant conditions is between 20 and 22°C. From Leinweber, 1956.

same is true of diurnal fluctuations in growth, turgor pressure, permeability, etc.

The very nature of the endogenous rhythm may be more closely approached when studying endogenous cycles of metabolism or enzyme activity (Venter, 1956). There are many papers dealing with these questions (for a discussion of this work see Bünning, 1956a).

In this respect the following point is particularly interesting. The endogenous periodicity is connected with cycles in the quantitative relation between the intensity of synthesizing and dissimilating processes. The capacities of synthesis and dissimilation change diurnally. The regulation of the endogenous periodicity by normal light-dark periods is of such a nature that the phase of high capacity for synthesis coincides more or less with the light period.

Clauss and Rau (1956) ascertained endogenous cycles of metabolic capacities by offering 72-hour light-dark cycles to young plants of *Hyoscyamus niger*. Each cycle consists of a 10-hour light period and a 62-hour dark period. It is sufficient to repeat the light periods every third day in order to regulate the endogenous rhythm and to allow its continuation during the long dark periods. Then an additional light period of 2 hours is given during the long dark periods. The length of this additional period is the same in each cycle, but its position within the cycle is different in the several experimental sets. The controls do not receive this light break. After the experiment is finished the quantity of chlorophyll produced by the several sets is measured. It becomes immediately evident that the capacity of the plant to synthesize chlorophyll with the help of the short additional light periods changes diurnally (Fig. 5). There are phases showing a very high capacity for this

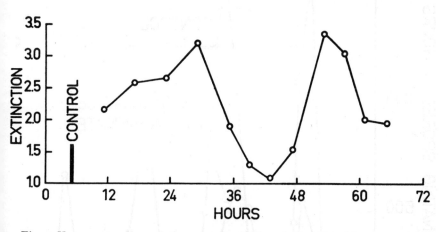

Fig. 5. *Hyoscyamus niger*, relative amounts of chlorophyll produced in 72-hour light-dark cycles. Each cycle consisted of 10 hours' light followed by 62 hours' darkness. During the long dark period an additional light period of 2 hours was given. The time of this additional light period was different in the several experimental sets. Controls without this light break. Abscissa: Time of the 2 hours light break within the 72-hour cycles. The curve shows that the light break influences chlorophyll formation in different degrees and even in a different direction at the several times. Phases with equal physiological character about 24 hours apart. From Clauss and Rau, 1956.

process, but other phases with a very low capacity. It even seems that during certain phases light decreases chlorophyll formation. Phases of equal physiological characteristics are about 24 hours apart. This is but

one example of many which illustrates the endogenous cycles of metabolic properties.

Metabolic cycles like these certainly are important for other diurnal
physiological processes governed by the endogenous periodicity. Several
years ago I tried to explain the mechanism of the endogenous rhythm
with the help of these metabolic cycles. But these and other attempts
failed, for the metabolic processes may be eliminated by enzyme poisons
without eliminating the rhythm. Bühnemann (1955a) in his experiments with *Oedogonium* applied poisons like NaCN, arsenate, 2.4-di-
nitrophenol, NaF, and others. Spore discharge in all these cases remains
a cyclical process as long as it continues (Fig. 6). Moreover, there is no
shifting of the phases and no increase in the length of the periods. Our
results with *Phaseolus* are similar (Bünning, 1956b; for 2.4-dinitro-
phenol compare Table V).

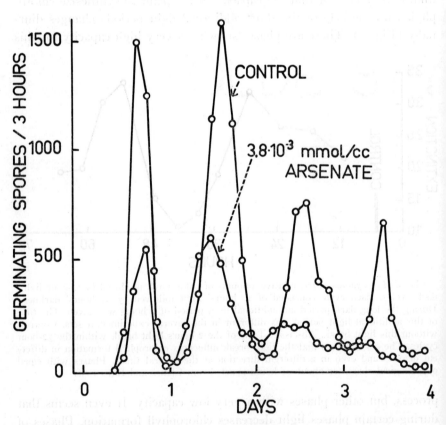

Fig. 6. Spore discharge in *Oedogonium cardiacum* in continuous light under the
influence of 3.8×10^{-3} mmol/cc $Na_2HAsO_4.7H_2O$. From Bühnemann, 1955a.

In these experiments one point is still very interesting. If these poisons have been active for several days and are then removed the rhythmical process starts once again. And, most interesting, there is no shifting of the phases. Maxima and minima occur at the same time as with the controls (Fig. 7). That means the clock system is still going on while the physiological process normally governed by it is fully

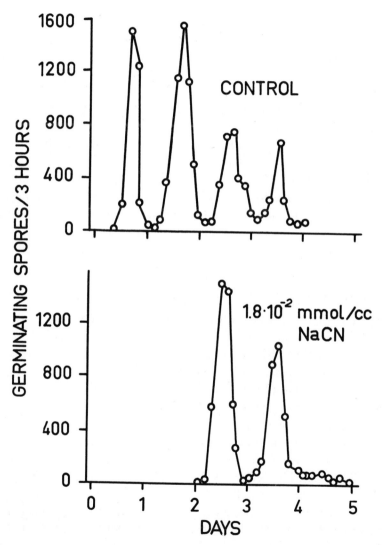

Fig. 7. Spore discharge in *Oedogonium cardiacum* in permanent light under the influence of 1.8 × 10⁻² mmol/cc NaCN. While the concentration of the poison decreases gradually, the periodicity starts again without a shifting of the phases. From Bühnemann, 1955a.

inhibited by poisons. We may draw the conclusion that perhaps physical processes are more important in the mechanism of the endogenous rhythm than chemical processes.

VII. RELATION TO THE STATUS OF PROTOPLASM

If periodic changes in the physical status of protoplasm are involved in the mechanism of the endogenous rhythm it should be possible to disturb the clock system by influencing the protoplasmic structure. This was attempted by several methods (Bünning, 1956b).

One of these methods was to offer the plant varying ratios of Ca:K. But this treatment did not have any influence on the endodiurnal system.

When plants are cut off and placed in solutions of different pH the rhythmicity disappears under extreme conditions. Especially in solutions with high pH values the rhythmicity becomes very irregular, and finally disappears (Fig. 8B). But this happens only at pH values which cause visible injury to the plant.

Later we tried to disturb the endogenous rhythm by exposing the plant for a short time to extremely high temperatures. After heating *Phaseolus* plants for 5 minutes to 80°C most of the leaf cells and petioles were dead. Only in the pulvini did living cells survive. After this treatment the periodicity is still visible, but the amplitudes are reduced. The length of the periods is unchanged.

VIII. EFFECT OF MITOTIC POISONS AND ANESTHETICS

Finally we found a chemical which is able to interfere with the working of the endodiurnal system without killing the cells (Bünning, 1956b, 1957). By applying colchicine to *Phaseolus* via the transpiration stream there occurs a distinct retardation of the rhythmicity. The periods can be extended to about 35 hours. In higher concentrations the periodicity disappears. But other types of influences on the rhythm may also be observed. It can become quite irregular (Fig. 8c). These results stimulated us to test other mitotic poisons. There are distinct effects of substances like urethane (Fig. 8D), trypaflavine (Fig. 8E), and acridineorange. The disturbing effect of these substances may easily be distinguished from that of metabolic poisons. As already mentioned metabolic poisons can gradually suppress the periodicity, but they do not make it irregular or shift the phases. Substances like colchicine, trypaflavine, acridineorange, on the other hand, cause specific irregularities of the periodicity before fully suppressing it. There are similar effects of ether (Fig. 8F). But the strongest effect was observed with phenylurethane.

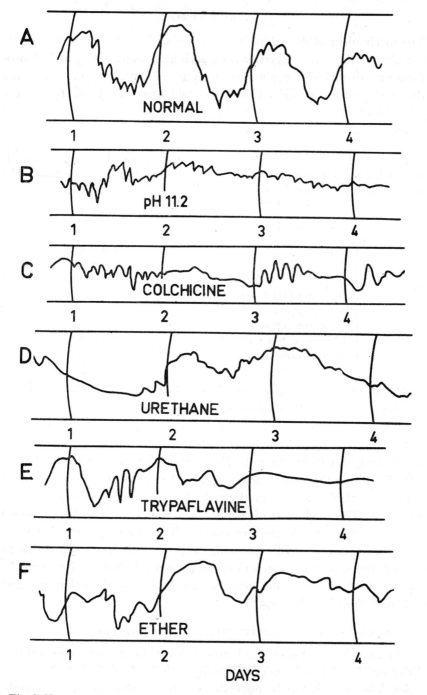

Fig. 8. Kymograph records of leaf movements of *Phaseolus multiflorus* in continuous light of low intensity, influenced by different chemical conditions. A. Controls in tap water. B. Leaves in water of *p*H 11.2. C. Leaves in water containing 0.4% colchicine. D. Leaves in water containing 0.5% urethane. E. Leaves in water containing 0.05% trypaflavine. F. Leaves in water containing 1.6% ether.

The mechanism of disturbing the rhythm with chemicals is apparently not always the same. Trypaflavine, acridineorange, ether, and chloroform may disturb the regularity of the curves. But they do not influence the length of the periods. Colchicine, and especially urethane, on the other hand, have also a distinct influence on the length of the periods (Table V).

TABLE V. *Phaseolus multiflorus*. Leaf movements under different chemical conditions. The poisons are offered with the transpiration stream. The concentrations are just below those which inhibit the movements.

Condition	Length of Periods (hours)
normal	25.0 ± 0.3
$pH = 3.2–4.0$	25.1 ± 0.7
$pH = 8.0$	24.1 ± 0.9
2.4-dinitrophenol 0.01%	26.1 ± 0.4
ether 4.5–7%	25.7 ± 0.8
chloroform 0.7%	27.5 ± 0.6
colchicine 0.3–0.4%	28.3 ± 0.6
urethane 1.0%	36.3 ± 1.4

The results are somewhat surprising since mitotic processes are not in all cases involved in the periodicity. The leaf movements for example are due to processes in cells which no longer show any cell divisions.

It seems that the mitotic cycle itself is a consequence of cyclic structural changes within the cell. These cycles are still going on while mitosis, due to the ageing of the cell, no longer occurs. The structural cycles reveal themselves in diurnal changes of the nuclear volume (Fig. 9). The beginning of a mitosis is well known to be connected with an increase in the nuclear volume. Now we may state that these volume changes are still continuing while mitosis has stopped due to ageing. But other consequences of the structural changes, for example, the periodicity in enzyme activity, are still possible (Bünning and Schöne-Schneiderhöhn, 1957).

IX. CYCLES OF SENSITIVITY TO EXTERNAL FACTORS

The two phases of about 12 hours each within the endogenous periodicity are known to be very different in their physiological characteristics. Therefore it can be understood that the plant reacts differently to the same external factors in the two different phases. One example of these cycles of sensitivity is the different amount of chlorophyll pro-

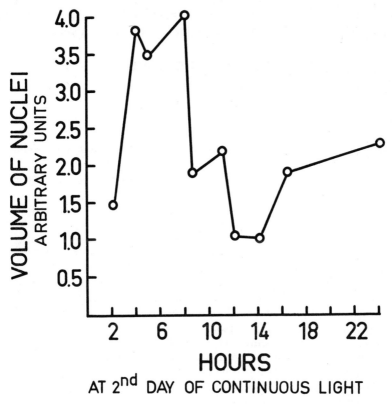

Fig. 9. Diurnal changes in the volume of nuclei (medium values) in guard cells of *Allium cepa*. Second day of continuous light. One unit = 1430μ.[3]

duced by a light break given at different times within a long dark period. Other very important examples are the photoperiodic and thermoperiodic reactions. It is my opinion that the photoperiodic processes induced by light or dark periods are nothing else but phases of the endogenous rhythm which is regulated by light and dark periods.

I do not intend to discuss this in detail, but I may mention that we published several experiments clearly showing diurnal cycles in the reaction to external factors like light and high or low temperature. These experiments were of the same type as the experiment on chlorophyll formation already described. During a long dark period the plant was offered light breaks at different times. In other experiments it was subjected to short periods of extreme low temperature. The reactions (for instance, flower formation, elongation, and other growth processes) were quite different depending on the time at which the plant was exposed to these special conditions of light or temperature.

It is this role of the endogenous rhythm in photoperiodism and thermoperiodism which gives the endodiurnal system a selective value.

BIBLIOGRAPHY

Ball, N. G., and I. J. Dyke. 1954. An endogenous 24-hour rhythm in the growth rate of the *Avena* coleoptile. *J. Exp. Bot. 5*, 421–433.

Bühnemann, F. 1955a. Das endodiurnale System der *Oedogonium*-Zelle. II. *Biol. Zentralbl. 74*, 691–705.

Bühnemann, F. 1955b. Das endodiurnale System der *Oedogonium*-Zelle. III. *Z.f. Naturforsch. 10b*, 305–312.

Bühnemann, F. 1955c. Das endodiurnale System der *Oedogonium*-Zelle. IV. *Planta 46*, 227–255.

Bünning, E. 1931. Untersuchungen über die autonomen tagesperiodischen Bewegungen der Primärblätter von *Phaseolus multiflorus*. *Jahrb.f. wiss.Bot. 75*, 439–480.

Bünning, E. 1956a. Endogenous rhythms in plants. *Ann. Rev. Plant Physiol. 7*, 71–90.

Bünning, E. 1956b. Versuche zur Beeinflussung der endogenen Tages-rhythmik durch chemische Faktoren. *Z.f.Bot. 44*, 515–529.

Bünning, E. 1957. Über die Urethanvergiftung der endogenen Tages-rhythmik. *Planta 48*, 453–458.

Bünning, E., and F. J. Leinweber. 1956. Die Korrektion des Temperatur-fehlers der endogenen Tagesrhythmik. *Naturwiss. 43*, 42.

Bünning, E., and G. Schöne-Schneiderhöhn. 1957. Die Bedeutung der Zellkerne im Mechanismus der endogenen Tagesrhythmik. *Planta 48*, 459–467.

Clauss, H., and W. Rau. 1956. Über die Blütenbildung von *Hyoscyamus niger* und *Arabidopsis thaliana* in 72-Stunden-Zyklen. *Z.f.Bot. 44*, 437–454.

Enderle, W. 1951. Tagesperiodische Wachstums-und Turgor-Schwankungen an Gewebekulturen. *Planta 39*, 570–588.

Flügel, A. 1949. Die Gesetzmässigkeiten der endogenen Tagesrhythmik. *Planta 37*, 337–375.

Leinweber, F. J. 1956. Über die Temperaturabhängigkeit der Periodenlänge bei der endogenen Tagesrhythmik. *Zeitschr.f.Bot. 44*, 337–364.

Pfeffer, W. 1907. Untersuchungen über die Entstehung der Schlafbewegun-gen der Blattorgane. *Abh.math.phys. Kl. Kgl. Sächs.Ges.d.Wiss. Leipzig 30*, 259–472.

Schmidle, A. 1951. Die Tagesperiodizität der asexuellen Reproduktion von *Pilobolus sphaerosporus*. *Arch.f.Mikrobiol. 16*, 80–100.

Venter, J. 1956. Untersuchungen über tagesperiodische Amylaseaktivitäts-schwankungen. *Zeitschr.f.Bot. 44*, 59–76.

Zinn, J. G. 1759. Von dem Schlafe der Pflanzen. *Hamburg. Magazin 22*, 40–50.

VII. PHOTOSYNTHESIS AND THE ORIGIN OF LIFE

BY H. GAFFRON[1]

I. PHOTOSYNTHESIS, A VERY IMPROBABLE REACTION

THE Darwinian theory of evolution has in the last two decades received strong support from a new side, the biochemistry of the living cell. The fundamental metabolic reactions which maintain life have been found to be of the same pattern everywhere and bound to the same few active substances which govern in each cell the supply of nutrients and of free energy. The enormous variety of more specific chemical reactions encountered in the world of living things all depends on a first set of basic organic reactions which are apparently needed to keep a cell alive. We biochemists are, of course, still far from being able to enumerate and describe the minimum number of reactions, and the simplest possible form of a metabolic mechanism sufficient to build and maintain "the aboriginal cell," the starting point of Darwinian evolution. Not long ago such a concept of a "minimum cell" would have been thought of as plainly absurd. Today it is quite a permissible abstraction indicating the goal of rather likely future achievements. And it has already encouraged further speculation reaching backward beyond the abstract model of "a living cell," into the time when there was no cell, only the material and the circumstances which allowed it to be created.

To the disgust of most scholars raised in the humanist tradition, many biologists today regard it not as a useful hypothesis, but as self-evident that living things must have evolved gradually from nonliving ones. They will not be deterred from asking questions about the origin of life merely because what we mean by life cannot be unequivocally defined in terms of our present knowledge and language, or because in the future they may come upon a barrier to further meaningful inquiry, similar to the one the physicists have met with in the realm of elementary particles. In short, the question as to the origin of life means for them an invitation into a new and promising field of experimental research.

During the last thirty years quite a few articles have been written describing how organic matter could have rapidly accumulated on earth soon after its crust cooled down to tolerable temperatures (see Bibliography, A, for references). Such accumulation and concentration

---

[1] Department of Biochemistry (Fels Fund), University of Chicago.

of organic substances, perhaps rather localized, must be assumed, it would seem, lest the subsequent improbable reactions we have to postulate become even more improbable. In the course of time these speculative discussions have brought about a certain measure of agreement. Our present chemical and biochemical knowledge supports, for instance, the following ideas. The ultraviolet rays in the solar spectrum are now generally considered to have been the main source of energy for the initiation of organic evolution. In addition, electrical discharges in the atmosphere, or volcanic activity bringing forth strange chemical reactions as well as heat, can be resorted to as far as necessary. Conditions were anaerobic. As the result of simple spontaneous condensations of molecules containing one or two carbon atoms, large amounts of organic material accumulated in the primeval oceans and were concentrated locally here and there. Mineral and metallic catalysts were abundant. By the time the influence of ultraviolet radiation on the course of events had become insignificant because of the formation of an ozone layer in the upper atmosphere, enough free energy had been stored in the mass of organic molecules to keep evolution going until photosynthesis developed and free oxygen became available in quantity. This finally made the further evolution and present abundance of living beings on earth possible.

Somewhere along the way, two or three incredible things happened: First the emergence of self-duplicating highly complex units and, during this process, the discrimination in favor of only one set of optical isomeres from the pairs of equally probable asymmetrical organic molecules. This was followed by the functional association of many such units leading towards a hierarchy of interdependent reactions separated from the outer world by a cell wall. Improbable as those events are, evidently there has been enough time for them to happen.

There is disagreement as to how much improbability can be tolerated in inventing models for these major steps in organic evolution. This question becomes the more pressing the shorter the time space is during which these decisive events must have occurred. Of late there is a tendency to date the birthday of the first living cell further and further back in order to have time enough for all the complexities and ramifications of Darwinian evolution. The oldest remnants of plants which have been discovered reveal full-fledged differentiated organisms, plants whose ancestors must have solved, long before, the problems of fermentation and respiration and photosynthesis. In this respect, the most ancient organisms we know were apparently quite up to date. Of course, an extremely improbable event can happen any moment. Part of the fascination which the problem of the origin of life holds for us stems from the apparent necessity to believe in events which happened only once—tantamount to acts of special creation and therefore never to be observed in the laboratory. It would certainly be a triumph of science if it could be demonstrated convincingly that life must have arisen by a process which could occur hardly at all in the lifetime of any one of the planets which must accompany millions of stars in thousands of galaxies. This would convey upon the fact of our existence a significance reaching far beyond our earthly limits. Such a future finding seems, however, unlikely; for history has shown that with increasing knowledge our position in the universe is shifting farther and farther away from any imagined center of importance.

I think modern biologists will instinctively choose to believe the opposite: life is bound to appear anywhere in the cosmos as soon as conditions become favorable. This presupposes that each individual link in the chain of reactions leading to the familiar manifestation of life had a sufficiently high probability to happen while the favorable conditions lasted. For instance, the often discussed spontaneous condensation of numerous amino acids into the shape of a modern enzyme evidently does not belong among the sufficiently probable reactions, and should for this very reason be dismissed as a possible steppingstone in the course of evolution.

On the other hand, as the word evolution connotes, the process has been a one-way street. Every once in a while a corner has been passed beyond which there was no return, no new beginning. From among the events along the road which have imposed this irreversibility we shall mention the following: loss of hydrogen gas into outer space; end of ultraviolet penetration to the surface of the earth; appearance of free oxygen in large quantity; destruction of the original organic matter by oxidation and by the activity of the first organisms.

Assuming that man-made radioactivity would finally kill the very last living cell, the task of re-awakening life on earth would look entirely different from what it must have been several billion years ago. While we know the first try eventually did succeed, we have a right to be skeptical about the likelihood that it could succeed a second time. Originally there may have been several developmental lines leading to self-reproducing entities. How many these were and when they became "dead-end streets" we shall probably never know. At present life is sustained by the free energy absorbed in the pigment system of the green plants and to a minute degree by the oxidation of inorganic reductants in a few autotrophic bacteria. Once the cells capable of photosynthesis are dead the pigment not only bleaches in the light but helps to oxidize other organic matter in the process. Similarly oxidation reactions are all that remain of the metabolic activity in the chemotrophic organism. A *de novo* evolution of life in our rather late era would mean that this trend towards complete oxidation would have to be reversed against heavy odds, while under the anaerobic conditions, in the early morning of evolution, this particular difficulty did not exist.

Seen from this angle, from the point of view of thermodynamics, photosynthesis is the most improbable metabolic reaction on earth. This became pointedly clear after it was established that the process works with 35% efficiency and yet is based upon the breaking of a bond in the molecule of water: something which needs much more energy than is available in a chlorophyll molecule after it has absorbed one quantum of visible light. No wonder that it is unique—and unique in more than one aspect. For instance, there are very many light-absorbing substances found in living cells, yet on this earth only one pigment, chlorophyll, a heterocyclic ring system containing four pyrrols and one five carbon ring (see Fig. 1), is indispensable for the task of converting the energy of light into chemical free energy.

Thus the life-sustaining function of chlorophyll seems to point backwards, just like the optical activity, to one singular event, a unique invention of nature. After many decades of research we have learned recently that this achievement is not due to the photochemical activity of chlorophyll alone. This astounding and highly improbable result comes about only in cooperation with other rather complex metabolic reactions, which in turn have been seen to occur independently in living cells devoid of chlorophyll and showing no photosynthesis of any kind. In other words, we have here a directed, "purposeful" interplay of several reactions, a system which, like the living cell itself, must have had an evolution and a history.

This history constitutes, of course, only one thread within the whole tissue of organic evolution. But I have the impression that because of various recent discoveries it has become possible to follow this thread backwards beyond the time of the early living cells—perhaps back to the very beginning of organic evolution.

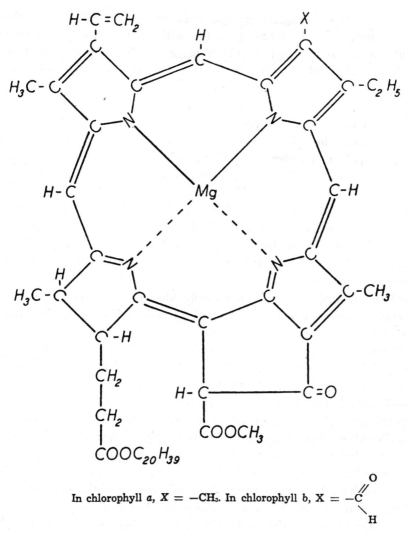

In chlorophyll *a*, X = —CH₃. In chlorophyll *b*, X = —C

Fig. 1. Structure of chlorophyll *a*.

II. THE EVOLUTION OF PHOTOSYNTHESIS

A. Experiments with ultraviolet. Synthesis of aliphatic acids and amino acids. Because all organic and living matter we see around us has been built up from carbon dioxide as the ultimate carbon source by means of photosynthesis, it seemed natural in the past to assume that CO_2 must also have been the carbon source at the time when organic compounds first appeared on the surface of the earth. For the same reason, namely, that photosynthesis was a photochemical reaction involving carbon dioxide, numerous experiments have been performed in the course of at least half a century with the expectation that either visible or ultraviolet light should transform carbon dioxide into something organic while dissolved in various colored or colorless mixtures.

Certain of these experiments met indeed with moderate success. In mixtures containing hydrogen and carbon dioxide, ultraviolet was shown to produce simple compounds like methane, formaldehyde, or formic acid. The effects of electrical discharges were similar. Since formaldehyde is a logical reduction product for a photochemistry involving the carbon dioxide molecule, this start of organic evolution seemed plausible. The more biochemistry advanced, however, and the more we learned about the mechanism of photosynthesis the less adequate any evolutionary hypothesis looked in which free carbon dioxide played the role of a major reagent. In photosynthesis it is water, not carbon dioxide, which participates in the reaction of chlorophyll. And among the numerous metabolic reactions observed in living cells, those involving formaldehyde are practically absent, while those centering around pyruvic, glyceric, acetic, and succinic acids and their amino derivatives predominate everywhere.

Meanwhile astronomers and geochemists debated about the likely composition of the early atmosphere on earth. At first, this discussion went on quite independently of the question of its possible relation to photosynthetic reactions. Urey (1952) was among those who believed that the atmosphere of the earth in its earliest stages contained no carbon dioxide because it was a highly reducing atmosphere. Specifically, he states that it must have been composed of hydrogen, methane, ammonia, and water vapor, with hydrogen in excess. Thus carbon dioxide could not have been the source of organic carbon if the formation of substances more complex than methane had begun already at that period in the history of the earth. Urey's atmosphere is colorless to our eyes; visible light has no power to change it. On the other hand, all the molecules present in the gas mixture decompose under the impact of short wave ultraviolet light. About 5% of the energy of the solar radiation reaching the earth is of this kind. In our time this part of the solar radiation no longer reaches the surface of the earth any more nor does it affect the composition of the air around us, because it is stopped by the ozone layer. But within a gas mixture of the sort we just mentioned, ultraviolet is bound to cause profound alterations. It is therefore likely that the world of organic substances started by means of photochemical reactions in which methane served as the original carbon source. Any other force creating broken molecules, that is, radicals, which can recombine at random, should give the same or very similar results: for example, electrical discharges or very high temperatures. But heat is likely to decompose the more complex molecules, the less probable combinations, faster than they are formed. For this reason, Urey's gas mixture cannot be changed by temperature alone.

On the other hand, radicals which are formed by radiation, by the direct absorption of single quanta, can recombine while in a cool environment in many ways and the resulting new compound may be quite stable at that particular temperature. It will thus tend to accumulate. Its light absorption spectrum, however, is different from that of its constituents; therefore, the new compound will now in turn be subjected to the impact of light of other wavelengths. This may lead either to destruction or to a further chain of new synthetic reactions. Harold Urey persuaded Dr. S. Miller (1955) to do a few experiments to see what happens in an atmosphere of this kind above a reservoir of boiling water under the influence of electrical discharges. The result turned out to be a very exciting one for all biochemists. The compounds formed were aliphatic acids, such as formic, acetic propionic, lactic and glycolic acids, and a number of amino acids, and some unidentified colored polymerized material. It is important that acetic acid, CH_3COOH, and glycine, HN_2CH_2COOH, were found in large amounts.

None of these compounds would have been synthesized in the presence of much oxygen. But some oxygen stemming from the decomposition of water must have been formed because of the carboxyl groups in the organic acid and the presence of a little carbon monoxide and carbon dioxide after the experiment.

This experiment, which has been independently confirmed, makes it extremely probable that such were the conditions and circumstances which started organic evolution once the temperature of the earth's surface had gone below the boiling point of water. The condensation of water in liquid form (the appearance of the first oceans) led to the preservation of the organic compounds after they had been formed in the gas or in the upper layers of water. They were dissolved in the depth where destructive radiation could no longer reach them.

Enormous amounts of organic material may have been formed this way. In the meantime hydrogen gas continued to escape into outer space. The atmosphere changed its composition to the point where some oxygen, formed by decomposition of water molecules, gave rise to photochemically formed ozone, and the latter began to shut off the ultraviolet radiation which until then had been the source of energy promoting the reactions we have been speaking about. This had two consequences. On the one hand, no more of the original organic material, the aliphatic and amino acids could accumulate, but on the other hand, the stage was now set for the appearance of more and more complex organic structures to which irradiation with ultraviolet would have been fatal. We now know that ultraviolet light is one of the best means to sterilize virus and microorganisms or to inactivate enzymes.

As an aside I would like to point out how rich a nutrient medium for present-day microorganisms Miller's brew seems to be. It might be worthwhile to look for organisms which will grow in such a synthetic soup under strict anaerobic conditions. There have been speculations about the possibility of spores coming from outer space and infecting the earth. At that particular time they would have found good breeding grounds. Such speculations, however, are of little interest since they evade and side-step the main problem.

As already stated, the possibilities which many of the proposed stories have in common are: 1) organic compounds dissolved in large amounts in the primeval oceans, 2) anaerobic conditions, and 3) eventual protection from the destructive forces of short wave ultraviolet. The promotive role of solar radiation is assumed to have stopped here. Only much later, after the first organism had succeeded in elaborating our familiar process of photosynthesis with chlorophyll, did it appear again and become all important.

In the meantime, evolution is supposed to have rolled along entirely at the expense of the free energy contained in a great variety of organic compounds reacting spontaneously with each other or with the molecules of water wherever local conditions favored specific condensations, hydrations, dismutations, and oxido-reductions.

Whichever way we look at this phase of evolution, we ought to admit that it must have been a slow one. Even if we accept the argument that the same reactions which are so speedily accomplished in living cells by specific enzymes are, from the point of view of thermodynamics, allowed to proceed in the same direction uncatalyzed and that the constituents of living cells should have eventually appeared, survived, and multiplied by good chance, this does not solve the question of the time required. We shall leave aside here the familiar calculations showing that certain important reactions must have, but never could have, happened. Considering that the average enzyme speeds up a possible reaction by the factor of 10^6 and that Darwinian evolution, while making use of such enzymes, needed between one and two billion years, there is really not much time left for an elaborate game of chance encounters and the selection of the most successful combination, culminating in the creation of an organism—all this at reaction velocities about a million times slower than observed in living cells.

We should, therefore, consider the following four possibilities:

First, that ultraviolet light continued to push forward synthetic reactions until photosynthesis took over. This would mean that the ozone layer was a consequence of photosynthesis and not of an earlier reaction producing free oxygen.

Second, that a comparatively small number of organic compounds sufficed to put the first self-reproducing system into business. This, to me, seems probable because of the astonishingly few types of prosthetic groups nature uses everywhere to keep the cell machinery running. The vast majority of man-made chemicals described in Beilstein's awesome handbook are not found in nature and, in addition, are mostly toxic for living organisms.

Third, that effective organic catalysts containing metal ions but of a structure much simpler than enzymes appeared rather early.

Fourth, that light did not stop having an accelerating influence at any time during the course of evolution because visible light did come into play with the aid of colored substances long before photosynthesis with chlorophyll was fully developed. As we shall see, new experiments on the natural synthesis of porphyrins lend some credibility to the last two propositions.

B. Formation of porphyrins and the possible role of light sensitized reactions. One of the impressive successes due to the radioactive tracer method in biochemistry has been the use of carbon 14 to elucidate the biosynthesis of porphyrins. Shemin, Rittenberg, and coworkers (see Bibliography, B) found in this way that acetate and glycine are used nearly exclusively by the living cell to construct such a complex molecular structure as the porphyrin ring. Furthermore, it apparently takes little time and energy to convert these simple two-carbon molecules into porphyrins. The steps by which the cell achieves this are surprisingly straightforward. The first intermediate derived from the simple nutrients, so abundant in Miller's synthetic brew, is δ-amino-laevuli(ni)c acid, a five-carbon keto-amino acid, apparently the product of direct condensations. Two of these molecules now combine to give one pyrrol ring with three side chains (see Fig. 2). This precursor pyrrol called porphobilinogen condenses again with itself to give a ring of four pyr-

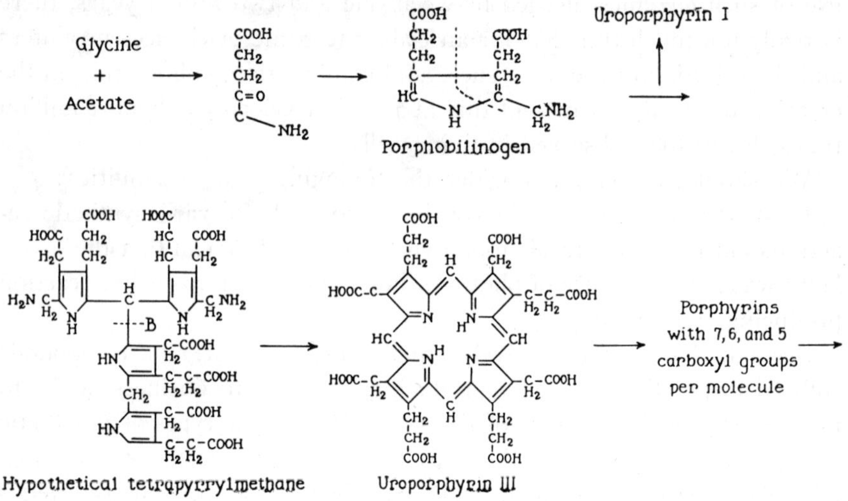

Fig. 2. Condensation reactions yielding porphobilinogen and formation of the porphyrin ring system.

rols, in other words a porphyrin, already with suitably reactive side chains attached. There seem to be intermediate condensation products having two and three pyrrols but the final stable end product from porphobilinogen is porphyrin. Parallel work by Della Rosa, Altman, and Salomon (1953), by Granick, Bogorad, and Jaffe (1953) and by Bogorad and Granick, 1953, emphasizes the fact that the same mechanism of porphyrin synthesis is found with little variation in bacteria, plants, and animals (Bibliography, B2).

As is well known, the porphyrins serve throughout the living world as the most ubiquitous reactive parts of enzymes. One or two iron-porphyrins are found in combination with a variety of proteins as oxidases, cytochromes, catalases, peroxidoses, and as constituents of the photoactive pigment system in purple bacteria and the green plants. In combination with cobalt, porphyrin appears in vitamin B_{12} which is indispensable not only for the health of the higher organisms, but also for the growth of certain algae. Finally, in combination with magnesium, porphyrin is found in two forms of chlorophyll. Thus porphyrin belongs to the small group of substances which like the pyrimidines, purines, and flavins sustain the basic reactions of living cells everywhere. If the first cell had any resemblance to what we now consider to be a living unit—and the Darwinian theory of evolution definitely requires such an assumption—the conclusion is inevitable that porphyrins must have played an essential role in the very first organisms. The identity of the physiology underlying all life points to the fact that as soon as a successful quasi-living mechanism appeared it conquered the earth, and no other independent evolutionary line has had a chance to survive. In order to envisage a gradual transition from the nonliving to the living, we have to assume that conditions outside the first self-reproducing unit were at that time quite similar to those inside of what we now call living protoplasm. In other words, the synthetic principles employed by nature two billion years ago should not have been much different from those which can still be seen at work in the majority of living cells. What amazes us is the rapidity as well as the precision with which these mechanisms now fulfill exactly what we are in the habit of calling their "purpose." But this problem can perhaps be explained away by the general notion of survival of the fittest. Once a certain metal-porphyrin attached itself to a protein molecule and thus reacted faster than before, working in seconds instead of a hundred years, it is self-evident that the protein enzyme complex had an enormous advantage over the catalysis by the prosthetic group alone (cf. Calvin, 1956). The same holds for specificity. A less refined catalytic system leads to the same results only more slowly. The specificity of enzymes can be compared to the pins coming out of a radio tube which have different thicknesses and different spacing so that they fit into the socket only in one position—the one which is functional. But this is merely an additional convenience for quick assembly. In principle, a number of pins which are indistinguishable would serve equally well except for the time it takes to try out and find the right position on the socket. Thus the problems of catalysis and of specificity as we see them today can be left aside for a while, and the important remaining question is whether such prosthetic groups, capable of speeding up a particular reaction, had a chance to be formed spontaneously. The answer is definitely yes—and in virtually no time at all.

The principle that the results of enzymatic reactions should be achieved without catalysis, provided we wait long enough, has been invoked to explain the spontaneous formation of the most marvelous structures. By contrast the condensation of acetic acid molecules and glycine into succinic and δ-aminolevulinic acids is certainly one of the easiest problems for nature to solve. The spontaneous condensation between two of the latter to form a substituted pyrrol ring has already been observed in the laboratory (Berlin et al., 1956). The resulting porphobilinogen is known to condense spontaneously to give porphyrin rings, particularly under oxidizing conditions or in the light. Thus with three steps starting from the level of acetic acid we arrive at a strongly fluorescent dyestuff—one of the most stable configurations in the organic chemistry of heterocyclic molecules. Porphyrins can stand a temperature of about 400°C before they disintegrate. They are resistant to oxidation and equally stable against ultraviolet light, at least as compared with the more sensitive compounds of the protein type. Considering that all kinds of metals were available on the surface of the earth at a time when there was a great accumulation of acetate and glycine due to the Urey-Miller process, the formation of porphyrins and metal-porphyrins must have been prompt and inevitable—in short, one of the earliest steps in evolution. No porphyrins have as yet been reported as products of the Miller experiment. This is no drawback at all for our hypothesis. It is amazing how well the Miller experiment succeeded, considering that a quartz vessel is so much plainer than the myriad different surfaces with which the first organic compounds came into contact. Under natural conditions a great number of diverse reactions were bound to follow, including the Shemin condensations. The high probability that porphyrins were formed in the initial stages of organic evolution fits, of course, well with the fact of their predominance as catalysts in present-day organism, as pointed out above.

We need not much imagination to see how important this evolutionary step has been. Not only was there immediately available an array of electron transfer catalysts for the interaction between molecules of different reduction levels, but also the most efficient kind of photocatalyst. With one stroke, as it were, at least one-third of the energy in the solar spectrum became available for the promotion of photochemical reactions.

By the time ultraviolet radiation stopped being effective on the surface of the earth, colored substances had taken over, supplying energy to the organic mixture obtained from sunlight. There may have been other active pigments besides the porphyrins. Certainly they did not survive in living systems. Experiments of the Miller type should provide us in the future with more solid knowledge in this respect. Of the porphyrins, we know that they can act as photocatalysts sensitizing hydrogen transfers between many different kinds of molecules. They must have, therefore, accelerated the interactions between the organic substances present at that time and thereby increased the diversity of carbon compounds on earth, and—because of this peculiarity of photoreactions, namely, to work very efficiently at low temperatures—produced substances which otherwise would not have had much of a chance for a continued existence.

Two other points are important. The light quanta absorbed by porphyrins are much smaller than those corresponding to the short wave ultraviolet (Fig. 3). As soon as the short wave radiation was stopped

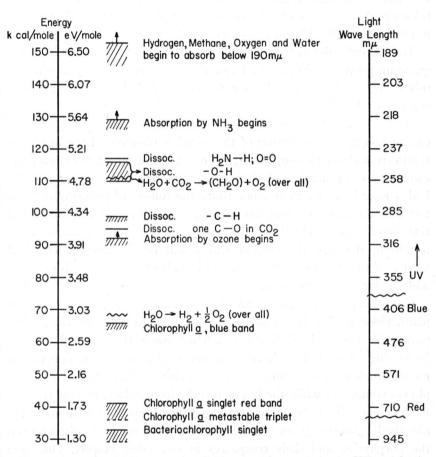

Fig. 3. Comparison of the energy available per mole quanta in the UV and in the visible light with that required for certain fundamental photochemical reactions. Note that the absorption by ozone in the upper atmosphere erases all possibilities of breaking an O—H, C—H, C—O, or N—H bond by means of a one quantum process at the surface of the earth. Substituting the immediate action of UV upon the molecules of methane, water, carbon dioxide, oxygen, etc., by chlorophyll-sensitized photochemistry requires the combined energies of several quanta of visible light. The energy available for chemical reactions by light-excited chlorophyll remains always below 41 kcal/mole. The finished over-all process of photosynthesis means a gain of 112 kcal. In the course of it many chemical bonds are broken and reformed; therefore, the actual energy to make the reaction go must obviously be much higher than the over-all gain. Measurements have shown that from six to ten light quanta are necessary.

by oxygen and ozone in the upper atmosphere, the danger was past that useful molecular combinations could be photochemically destroyed, or early organisms inactivated. By comparison, the photochemistry in visible light is known to be rather harmless, certainly as long as one very reactive element is absent—oxygen. Conditions, therefore, must still have been mainly anaerobic; for porphyrin-sensitized oxidations belong to the most efficient photoreactions studied and are as destructive in their way as the big quanta of ultraviolet light.

The second point is that these reactions must have been at best one quantum processes, just as we know them to proceed easily in our test tubes. As we shall see, this is a fundamental difference from photosynthesis.

H. F. Blum (Bibliography, A2) has already pointed out that we cannot imagine the evolution of the complex photochemistry in photosynthesis without passing through the one quantum stage of reactions common to any number of chemically unrelated fluorescent dyestuffs. I also would like to quote from Blum the following paragraph: "The fact that the same dye may act as a light absorber for . . . dissimilar reactions suggests that the environment . . . is more important in determining the type of reaction which will occur than the chemical nature of the dye molecule itself." This lack of specificity disappeared when chlorophyll was formed.

C. Chlorophyll and its reactions with water. If extracted from the plant, chlorophyll behaves in the light essentially like any fluorescent porphyrin. The study of chlorophyll, which contains a magnesium atom in its ring where hemins have iron, has taught us that the lightweight metals, such as magnesium, zinc, and calcium, give photoactive porphyrin complexes while the heavy metals, iron, cobalt, or copper, give inactive ones. If the latter are irradiated they immediately dissipate the absorbed light energy as heat. The chlorophyll molecule differs from the porphyrins and their complexes in one other respect. One side chain in natural porphyrin has been bent around to form a fifth ring adjacent to one of the pyrrols (see Figs. 1 and 2). Studies on the evolution of chlorophyll in present-day plants by Granick, Bogorad, and others (Bibliography, B2), leave no doubt that the porphyrins are the evolutionary precursors to chlorophyll. With the addition of an atom of magnesium and the closing of a fifth ring, we obtain a catalyst which is, as far as we know, the only indispensable dyestuff participating chemically in the photosynthetic reaction. Again the transition from porphyrin to chlorophyll is of such a nature that it could have happened easily in the absence of any living cell.

As pointed out before, the essence of photosynthesis is not a photochemistry of the carbon dioxide molecule, but one involving the molecules of water. This fifth ring seems to be the very instrument by which water can be photochemically separated into its components.

As shown in the scheme of Fig. 4, which has been drawn according

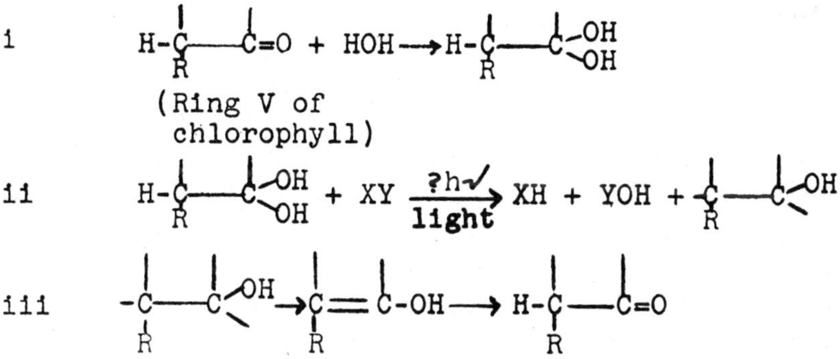

Fig. 4. Scheme showing how a water molecule may be forced apart photochemically when attached to a chlorophyll molecule (cf. Franck, 1955).

to ideas lately revived by J. Franck (1955), water enters the molecule by enolizing the keto group in this fifth ring. Subsequently the water molecule is removed by a succession of photo and coupled chemical reactions which together constitute a small metabolic cycle leading to the separation of one hydrogen from one hydroxyl. Nothing forces us to assume that an entire living cell is necessary to make this reaction occur. Indeed, we know this not to be the case and the question is only how much of the parts of a surviving chloroplast (after the cell has been broken) can be removed without interfering with the photodissociation of water molecules with the aid of visible light. For the first time in the course of this story we now come to a point where we are left without any supporting facts. I choose to believe that this interaction between chlorophyll and water also developed spontaneously perhaps with the aid of a solid-liquid interface, as the third important step in the evolution of photosynthesis. All that seems necessary, as far as our knowledge of photosynthesis goes, is a certain structural arrangement of several hundred molecules, and fitting reagents (xy) to take over, on the one hand, the hydrogen and, on the other, the hydroxyl from water.

$$\text{XY} + [\text{HOH, Chloroph.}] \xrightarrow{\text{light}} \text{XH} + \text{YOH} + \text{Chlorophyll}$$

Written as an over-all process, this looks very akin to an ordinary hydrolysis. Still behind this hides a crucial problem.

The immediate products of the photochemistry contain sufficient energy to promote a sequence of reactions which in the green plants leads to a saving of 35% of the absorbed light energy in the form of stored chemicals. Under these circumstances a one quantum process per water molecule decomposed is impossible (see Fig. 3). Direct measurements show that about two quanta have to cooperate to pry apart the bond between H and OH. How this is done we do not know. There exist many theories regarding this effect. It is possible that the structural arrangement of the chlorophyll molecules permits a "double excitation." One thing is certain: whatever the solution of this key problem, there will be nothing mystical about it—no "life force" need be invoked. Let us go on with the story.

As it should be in organic evolution, the successive steps lead to ever more complicated arrangements occurring less and less frequently. There is the fact that chlorophyll (existing in three very closely related variations) is one and everywhere the same! Is this so because, in spite of appearances, the precursors, the porphyrins, and chlorophyll itself could be formed only by an extraordinary stroke of luck, or because the two-quantum step was the rare achievement and the consequences of the reaction with water were so far-reaching that no other similar process appearing later had any chance to compete successfully against it? I believe the cooperation between two light-absorbing acts to be the unique trick. All other photochemical reactions in living cells seem to be of the ordinary one-quantum type.

Granick has already pointed out that present-day intermediates in the evolution of chlorophyll were at some time at the end of the chain, and probably had some function.

One example showing that light may sometimes be useful yet not absolutely necessary, is the elaboration of pigments in plants. The green color appears in seedlings of higher plants only under the influence of light, while many algae develop their chlorophyll in complete darkness. Thus the higher plants apparently have lost one link in the enzymatic chain leading to chlorophyll. They have replaced the action of an accelerating catalyst by a photocatalysis of the kind we have attributed to the early porphyrins. Where the collecting of energy is not of the essence, where a trigger action of light—as in vision—is all that is required, a simple photoactivation with one quantum is sufficient. Consequently there exist in the living world several such light-catalysed reactions which have little or nothing in common with photosynthesis. This view also agrees very well with the occurrence of a variety of other pigments which are found associated here and there with the photosynthetic apparatus. They are the phycobilins, fucoxanthins, and various carotenoids. Their distribution among different species of plants is very uneven, and it has been shown definitely that they contribute to photosynthesis as an accessory for better absorption of visible light. The energy of the light they absorb is immediately and directly transferred to chlorophyll *a*. In the course of evolution, these numerous other chloroplast pigments, of which there are more than forty, have probably arisen secondarily. Strain (Bibliography, c3) says that the great variation in the amounts of phycobilins in closely related organisms indicates that these pigments may never have been indispensable constituents of the photosynthetic apparatus and that their significance in the phylogenetic relationship is not clear.

For the benefit of those who, not being specialists, rightly associate with the word photosynthesis the assimilation of carbon and the evolution of oxygen, it ought to be mentioned here that both these reactions can be experimentally dissociated from the water-splitting capacity of the chloroplast. In other words, the complete photosynthetic system consists of several semi-independent sets of reactions, and it seems to me there is no other way to account for this from the point of view of evolution, than to let the photochemistry first develop to a level where the other partial reactions can be added and fitted together logically and without effort.

If we now have, thanks to the fifth ring in chlorophyll, a photoreaction in which water participates, we can expect a number of new reactions to happen. Let us condense the scheme of Fig. 4 and symbolize the photochemical reaction by this simple equation:

(a) $$n \cdot hr + [HOH] \longrightarrow [H] + [OH]$$

[H] and [OH] contain that part of the light energy which was not unavoidably lost in the process of their formation. The natural tendency of this pair of reducing and oxidizing substances (radicals?) is, of course, to recombine to water. And we have evidence that in photosynthetic organisms such immediate recombination takes place. Obviously, in the normal course of events the recombination or back reaction is successfully avoided. The usual way of the living cell to minimize back reactions following endergonic reactions where precious free energy has been invested to push the reaction forward, is to couple the initial reaction with secondary ones which remove the products of the first before they can get lost by recombination. The same principle will serve us here. We assume the photochemical act is followed by other reactions which quickly engage [H] or [OH] or both. To find a model and support for what I like to regard as the simplest type of photochemical reaction involving water, we turn to certain species among red-colored photoheterotrophic organisms called purple bacteria. Unlike the green plants they never release free oxygen in the light. Yet they depend for all their metabolic activities on light as the driving agent. They cannot respire—they are not equipped for it—and die if exposed to oxygen. They do not ferment either; at least what little fermentation has been found in Rhodovibrio is incapable of furnishing enough energy to sustain further growth. But in the presence of acetic acid, and nothing but acetic acid (except the usual mineral nutrients), these bacteria multiply vigorously in the light. All the cell constituents are derived from the one material which appears in abundance in the Miller-Urey reaction. How this can be brought about by the photochemical splitting of water is indicated in the scheme of Fig. 5. We need only to postulate that acetic acid first condenses to aceto-acetic acid. One-half of this four-carbon compound becomes reduced by the

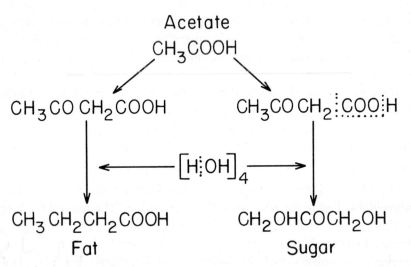

Fig. 5. Scheme showing how acetate may be converted into cell material by the photochemical reaction with water molecules in certain purple bacteria using infrared light.

hydrogen [H] from water; the other half oxidized by the [OH] group, and the result is, on the one hand, the production of fats, on the other hand, the production of carbohydrates with the release of some carbon dioxide. The over-all process, as well as most of the required intermediate reactions, have indeed been observed in purple bacteria. The single steps as shown here are not meant to be correct or complete in their biochemical details. These vary from species to species. The principle is what matters. Such a photochemical mechanism enables the bacteria to grow in the light at the expense of one organic compound. After the transformation of acetic acid into bacteria, they themselves, considered as another organic compound, need not have a greater free energy content than the acetic acid from which they have been formed. The light absorbed by chlorophyll (bacterio-chlorophyll is slightly different from plant chlorophyll) is used here to produce *different* and complex organic substances, not *more* of them. As we shall see, the creation of a larger amount of organic matter than was previously available requires additional steps.

In this way light and a pigment such as bacterio-chlorophyll may have been essential for the speedy evolution of the first organism as long as there was plenty of reducing material around. The photoreactions involving water promoted certain reactions which under anaerobic conditions would hardly have happened spontaneously in the dark.

Neither the energy originally stored in the organic substances by the Urey-Miller reaction, nor the total amount need to have changed much during this phase of evolution.

D. Photoreduction. The same bacteria, or at least closely related ones, can make use of either hydrogen sulfide or of molecular hydrogen to fix free carbon dioxide in the light. In this case there is, of course, an increase in organic substance. Since hydrogen and carbon dioxide are the simplest nutrients imaginable, such photo-autotrophic bacteria were considered for a while as the models for the earliest organisms. According to the line we have followed so far in this essay, it seems more logical to assume that the first light-dependent organisms lived and grew according to the principle discussed above. Later, they went one step ahead and learned, still under anaerobic conditions, to assimilate hydrogen, carbon dioxide, and even free nitrogen when the original organic material, Miller's acetic acid and glycine, became scarce.

This more advanced type of light metabolism is called "photoreduction" for short. What happens can be summarized using the symbols of equation (a) and $[CO_2]$ to indicate a bound form of carbon dioxide, as follows:

(b) $$[CO_2] + 4[H] \longrightarrow [CH_2O] + H_2O$$
organic

(c) $$4[OH] + 2H_2 \longrightarrow 4H_2O$$

The new twist is the assimilation of carbon dioxide, a very important step. On closer inspection, however, it turns out that as far as the role of chlorophyll is concerned photoreduction of carbon dioxide is hardly a more difficult achievement than the assimilation of acetic acid, in the scheme of Fig. 5. Instead of the keto group in aceto-acetic acid a carboxyl group is reduced to aldehyde, $RCOOH \longrightarrow RCHO$. And it so happens that this carboxyl group originates by means of a carboxylation reaction involving free carbon dioxide, $RH + CO_2 \longrightarrow RCOOH$.

Several such reactions in which carbon dioxide becomes an organic carboxyl group have been discovered in living cells. They occur independently of the presence or absence of a chlorophyll system, though all may become coupled to it. One of them, the formation of phosphorylated glyceric acid from CO_2 and ribulose diphosphate, appears to be particularly suited to make the most efficient use of the photochemistry in the chlorophyll complex. But even this carboxylation reaction is no exclusive property of photosynthetic organisms.

Reaction (b) above will not proceed, however, even if a carboxylating system is available unless reaction (c) proceeds simultaneously. In its simplest case (c) is a reduction by molecular hydrogen. Considering that the early atmosphere is supposed to have contained quite a bit of hydrogen it should not be too far-fetched to believe that organo-metallic catalysts made hydrogen available for all kinds of metabolic reactions. Hydrogenase, as this type of catalyst is called, occurs in a variety of microorganisms—again irrespective of whether they are photosynthetic organisms or not. The anaerobic purple bacteria are dependent for their photoreduction with hydrogen (or hydrogen sulfide) on the presence of hydrogenase just as they need a proper carboxylating system to incorporate carbon dioxide. Ever since this photoreduction

(d) $$2H_2R + CO_2 \xrightarrow{\text{light}} (CH_2O) + H_2O + 2R$$
organic

was discovered (van Niel, 1931) it has been a puzzle why so much light energy was needed to drive the reaction forward. Thermodynamically there is not much difference between the sum of the free energy on the left side and that on the right side of equation (d). One light quantum should be quite sufficient to achieve the reduction of one molecule of carbon dioxide, or even of many more. Yet direct measurements have shown that about six to nine quanta are required; the same as in the green plants, where the energy levels between the starting material and the products are 120 kcal apart. There was no satisfactory explanation other than to assume that photoreduction involved (quite unnecessarily it seemed) also a photodecomposition of water molecules as shown in reactions (b) and (c) and the scheme of Fig. 5. This makes sense only from the point of view of evolution. In this connection the question arises whether photoreduction and the metabolism of purple bacteria should be considered at all as an evolutionary step preceding photosynthesis in plants. It could just be a chance development branching off from the main evolutionary tree.

The following appears to me to speak against this and to support the story as given. Hydrogenase was found to be present in some strains of widely different species of algae: green, brown, blue-green, and red. The occurrence is completely random. Among closely related species which live and grow exactly in the same way, one may have it, the other not. Normally, the hydrogenase is completely inactive. Very little oxygen is sufficient to keep it so. These plants are aerobic organisms, do not grow under anaerobic conditions, and develop oxygen when illuminated—in short they have no use for hydrogenase. Yet when incubated a few hours in the presence of pure hydrogen gas they can perform nearly all the reactions with molecular hydrogen known to occur in typical hydrogen-using bacteria, with the difference that the bacteria can grow anaerobically in hydrogen while the algae can not. The result of this adaptation to hydrogen, as far as the light reactions are concerned, is that the plants stop evolving oxygen and take up the equivalent amount of hydrogen instead. The assimilation of carbon dioxide meanwhile continues undisturbed.

Horwitz and Allen (1957) could show that there is a competition in the green cell between the two forms of carbon dioxide reduction. Conditions that favor oxygen evolution hinder the working of the hydrogenase and vice versa. The changes are gradual and perfectly reversible so long as no oxygen is allowed to accumulate. These observations, it seems, provide us with an explanation of how the last evolutionary step leading to complete photosynthesis came about.

E. *Photosynthesis with the evolution of oxygen.* The organisms whose metabolic reactions we have so far taken as models explaining evolutionary patterns are dependent for their growth not only on light but also on the presence of substances which can serve as reducing agents. They are unicellular, anaerobic, of a reddish color, and incapable of evolving oxygen in the light. No multicellular, differentiated anaerobes are known. The plants, on the other hand, encompass an enormous range from the unicellular to the most differentiated multicellular forms. They grow in air, are easily damaged by anaerobic conditions, do not depend on reducing substances, but in their green parts assimilate carbon dioxide in the light while evolving oxygen. Quite a contrast; and yet, we contend, this enormous evolutionary step was initiated by two relatively small changes in the photochemical mechanism of the anaerobic organisms, namely, a shift in color and the substitution of reaction

(c) $$4[OH] + 2H_2 \longrightarrow 4H_2O$$

by

(e) $$4[OH] \longrightarrow 2H_2 + O_2$$

i.e. a dismutation in place of a reduction.

The change in color of the cells from purple to green is the visible sign for the fact that plant chlorophyll transforms more energy per light quantum absorbed than bacterio-chlorophyll. The purple bacteria make use of light quanta containing 30–35 kcal per einstein. The smallest light quantum usefully absorbed by the chlorophyll of green plants contains 41 kcal.

Reaction (c) should easily proceed with the release of energy even if [OH] were strongly bound to the substance acting as an intermediate.

The dismutation (e), by contrast, will not happen unless [OH] is a reactive group. This requires more energy. It seems, logical, therefore, to attribute the capacity of the green plant for releasing free oxygen at least partly to the somewhat greater energy available in plant chlorophyll. Yet this alone does not guarantee that reaction (e) really proceeds in the desired way. The existence of photoreduction in green algae emphasizes this point. As long as the [OH] is more efficiently reduced back to water than it can dismutate with other molecules of its own kind, there will be no evolution of oxygen. Therefore, the final step towards full photosynthesis required, besides an increase in energy, a weakening of coupling with the hydrogenase.

The competition between the two reactions (a) and (e) had to be weighted definitely in favor of (e). If later on the hydrogenase system could be put out of action entirely, photosynthesis as we know it would have been finally established. Fortunately, this end of the story is no mere speculation. There exists a special catalyst in the plants which promotes the release of free oxygen. The essential part of it has been recognized as a manganese ion (very likely bound to some protein). E. Kessler (1955) grew algae of the hydrogenase-containing varieties in absence of manganese. Such manganese deficiency was already (Pirson et al., 1952) known to decrease the rate of photosynthesis. It turned out that the more the capacity for evolving oxygen declined with a more severe manganese deficiency in these algae, the better became their ability to reduce carbon dioxide by photoreduction. The rates of hydrogen uptake in the light in manganese-deficient cells surpassed several times the rates ever measured before in normal algae. Normally grown hydrogen-adapted algae switch back easily to the evolution of oxygen if they are illuminated with strong light—very likely because now the concentration of [OH] favors the dismutation reaction. The manganese-deficient hydrogen-adapted algae respond to the same increase in light intensity only with a faster rate of photoreduction. Our present knowledge of the relationship between the two ways of carbon dioxide assimilation is schematically depicted in Fig. 6. Addition of the miss-

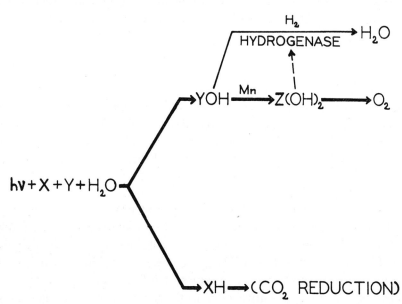

Fig. 6. Simple scheme for carbon dioxide assimilation showing the pathways of photoreduction and of photosynthesis in green cells containing hydrogenase. Aerobically only the way to oxygen evolution is open; hydrogenase is inactive. Anaerobically, hydrogenase functions. A competition ensues between oxygen evolution and reduction of [OH] again to water. In presence of specific poisons for the system leading to oxygen or in absence of manganese, only photoreduction takes place.

ing ions to the environment in which the manganese-deficient algae are being studied is all that is necessary to shift the photochemical metabolism again in favor of the evolution of oxygen.

If we transpose these observations into our story of evolution it is hard to believe that it was the manganese ion which came in as the very last piece to complete the machinery of photosynthesis. This metal must have been available all the time. Thus, the last step was either the change from bacterio-chlorophyll to plant chlorophyll with the concomitant increase in available energy or the appearance of the right protein which conveyed upon the manganese ion the proper catalytic power. Whichever step was *last*, the result was the end of the anaerobic era on earth.

The victory of the oxygen-resistant and oxygen-using organisms became universal after the partial pressure of oxygen rose high enough to oxidize the hydrogenase and similar anaerobic catalytic mechanisms into inactivity (Pasteur effect). Before becoming inactive by oxidation the hydrogenase in our green algae usually puts up a losing fight for survival, using the hydrogen to reduce the oxygen to water. This is a reaction which in other colorless microorganisms developed into their main source of energy for carbon dioxide reduction and growth. Strangely enough, this coupling with carbon dioxide reduction happens to be present also in our green algae where it leads only a potential existence and is of no use whatever. This I mention here only as one more support for the evolutionary point of view that the most successful and apparently most purposeful process to forestall the final decay of organic matter and of living organisms came about by the chance combination of several developmental lines in the organic world with a photochemical system whose beginnings probably antedated them all. (Fig. 7).

III. CONCLUSION

With the completion of a photosynthetic system using water and carbon dioxide as raw materials, while discarding oxygen as a waste product, began the second period during which solar energy was the driving force for the accumulation of large masses of organic matter—this time indirectly with the aid of enzymes in living cells. The success of this evolutionary step is entirely due to the fact that oxygen reacts only slowly with most organic substances at temperatures below 100°. But the perennial forest fires are also a consequence of photosynthesis because the plants have raised the oxygen content of the atmosphere to 21%, a partial pressure over a hundred times higher than necessary to main-

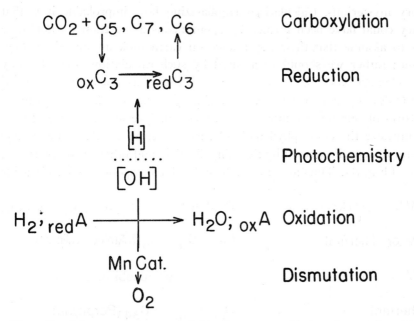

Fig. 7. The different components, probably of separate and independent origin, which together constitute photosynthesis or photoreduction in green plants and purple bacteria.

tain the full rate of respiration in most unicellular algae and other microorganisms. On the other hand, the content of carbon dioxide in the air has gone down to 0.03%—not enough to satisfy the photosynthetic reaction in bright daylight. This means that during the last millions of years photosynthesis has not any more materially increased the bulk of organic and living matter on earth. As the deposits of coal, oil, and of limestone of recent origin reveal, there may have been periods when the earth was richer in terms of total living matter.

The foregoing story, explaining the development of the mechanisms of photosynthesis as we now see it, is not entirely arbitrary. It contains sufficient factual elements to serve perhaps as a model showing how the evolution of other fundamental metabolic reaction might also become understandable as a sequence of rather simple steps. From the moment, however, that photochemistry joined hands with such originally independent metabolic systems as carboxylation and decomposition of peroxides, we can hardly afford to ignore the role of enzymes. There is no use in skipping lightly over the special difficulty of explaining the appearance of the duplicating catalytic proteins or nucleic acids. And as everybody knows, their evolution seems at this moment still very mysterious. Calculations emphasizing how improbable it is that they could have been formed by spontaneous association do not force us to assume that their appearance on earth took an incredibly long time; rather we should be assured by such calculations that the way the chemist has thought of synthesizing them has not been the way of their development in nature. How clumsy seem now the long arduous efforts of our most famous scientists to analyze and especially to re-synthesize the chlorophyll molecule, as compared with the few simple natural steps discovered by Shemin, Granick, and others (Bibliography, B2) (Fig. 8). Thus also the high molecular compounds (nucleic acids

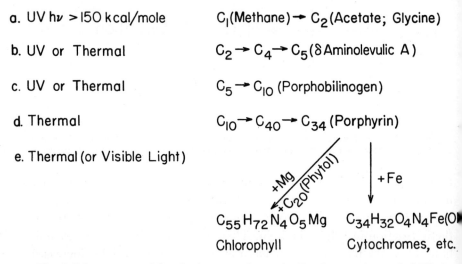

a. UV $h\nu > 150$ kcal/mole C_1(Methane) $\rightarrow$ C_2(Acetate; Glycine)

b. UV or Thermal $C_2 \rightarrow C_4 \rightarrow C_5$ (δ Aminolevulic A)

c. UV or Thermal $C_5 \rightarrow C_{10}$ (Porphobilinogen)

d. Thermal $C_{10} \rightarrow C_{40} \rightarrow C_{34}$ (Porphyrin)

e. Thermal (or Visible Light)

$+Mg$ $+C_{20}$(Phytol) $+Fe$

$C_{55}H_{72}N_4O_5Mg$ $C_{34}H_{32}O_4N_4Fe(O)$

Chlorophyll   Cytochromes, etc.

Fig. 8. Scheme summarizing the few natural steps leading from methane, via Miller's acetic acid and glycine, to photochemically active porphyrins, iron-porphyrin catalysts, and chlorophyll.

and proteins) may originally have arisen much faster than we are able to imagine at the present time (Fox et al., 1955). There is still a difference of several magnitudes in complexity, however, and this makes it nearly certain that porphyrins and chlorophylls were around a long time before the modern proteins made their appearance. On the other hand, astronomers and geophysicists have given us a starting date. We cannot retreat beyond that. Thus, we are not allowed to wait indefinitely for the most improbable event to happen. We must make an effort to confine ourselves to hypotheses where things proceed much faster than hitherto supposed. One way of speeding up evolution in such a picture is to supply a steady stream of free energy to take the place of ultra-violet light. Only dyes can catch the energy contained in daylight.

Among the colored organic and inorganic substances which may have played a role in the early stages of organic evolution, the porphyrins and chlorophyll survive as photocatalysts to this very day. This is perhaps because of their increased efficiency after they became part of protein complexes. In closing, I would like to point to the ease with which such complexes are formed once the right protein is present and the tremendous difference that a new combination may bring about in the fate of an organism. We have a suitable laboratory model in the work of Jensen and Thofern (1953). They obtained a strain of micrococcus which did not contain cytochromes. It was a strictly anaerobic organism, capable of growing well on suitable organic media provided oxygen was absent. In air the cultures died. But all that was needed to prevent this and to enable the bacteria to cope with the new situation was the addition of some ferri-porphyrin chloride to the nutrient medium. This the bacteria incorporated into their cells and as long as the supply of iron porphyrin lasted they continued to grow aerobically; for now they were in possession of a catalase and of a suitable respiratory system. Resting bacteria did as well in incorporating the iron porphyrin as growing ones. Interestingly enough, the resulting enzymes were spectroscopically and chemically different from those functioning in naturally occurring aerobic strains of the same micrococcus.

Something resembling this model of adaptation to new surroundings has very likely happened in the remote past—not once, but again and again—on successive levels of organization until the new faculties acquired by chance combinations became encased into a rigid genetic system of organisms which are no longer capable of transmitting acquired circumstantial habits to their offspring.

BIBLIOGRAPHY

The literature which could be quoted to sustain the argument in this article is tremendous. We have chosen, therefore, to list primarily reviews of the different fields in question, in addition to a few individual articles.

A. *Origin of Life*
    1. Blum, H. F. 1955. Perspectives in evolution. *American Scientist 43*, 595–610.
    2. Blum, H. F. 1955. *Time's Arrow and Evolution*, 2nd ed. Princeton Univ. Press, Princeton.
    3. J. B. S. Haldane, J. D. Bernal, N. W. Pirie, J. W. S. Pringle. 1954. Contributions in *The Origin of Life*, New Biology 16. Penguin Books, London.
    4. Oparin, A. J. 1938. *The Origin of Life*. Macmillan, New York.
    5. Wald, George. 1954. The origin of life. *Scientific American 191*, No. 2, August, 44–53.

B. *Porphyrin*
   1. Lemberg, R., and J. W. Legge. 1949. *Hematin Compounds and Bile Pigments*. Interscience, New York.
   2. *Symposium on Porphyrin Biosynthesis and Metabolism*. 1955. Ed. by G. E. W. Wolstenholme. Ciba Foundation. Contributions by D. Shemin and by S. Granick. Little, Brown, and Company, Boston.
C. *Photosynthesis*
   1. Larsen, H. 1954. The photolitho-autotrophic bacteria and their energy relations. In *Autotrophic Micro-organisms*, 4th Symposium of the Society for General Microbiology. Univ. Press, Cambridge, England.
   2. Rabinowitch, E. 1945, 1956. *Photosynthesis*, Vol. I, Vol. II. Interscience, New York.
   3. Strain, H. H. 1938. *Leaf xanthophylls*. Carnegie Institution. Publ. 490.

Berlin, N. I., A. Neuberger, and J. J. Scott. 1956. The metabolism of δ-aminolaevulic acid. I. Normal pathways, studied with the aid of ^{15}N. *Biochem. J. 64*, 80–90.

Bogorad, L., and S. Granick. 1953. The enzymatic synthesis of porphyrins from porphobilinogen. *Proc. Nat. Acad. Sci. 39*, 1176–1188.

Calvin, M. 1956. Chemical evolution and the origin of life. *American Scientist 44*, 248–263.

Della Rosa, R., K. I. Altman, and K. Salomon. 1953. The biosynthesis of chlorophyll as studied with labeled glycine and acetic acid. *J. Biol. Chem. 202*, 771–779.

Fox, S. W., D. DeFontaine, and P. G. Homeyer. 1955. Protein genealogy. *Fed. Proc. 14*, 213.

Franck, J. 1955. Physical problems of photosynthesis. *Proc. Amer. Acad. Arts & Sci. 86*, 17–42.

Granick, S., L. Bogorad, and H. Jaffe. 1953. Hematoporphyrin IX a probable precursor of protoporphyrin in the biosynthetic chain of heme and chlorophyll. *J. Biol. Chem. 202*, 801–813.

Horwitz, I., and F. L. Allen. 1957. Oxygen evolution and photoreduction in adapted *Scenedesmus*. *Arch. Biochem. Biophys. 66*, 45–63.

Kessler, E. 1955. On the role of manganese in the oxygen-evolving system of photosynthesis. *Arch. Biochem. Biophys. 59*, 527–529.

Jensen, J., and E. Thofern. 1953. Synthesis of the hematin enzyme. *Z. Naturforsch. 8b*, 604–607.

Miller, S. L. 1955. Production of some organic compounds under possible primitive earth conditions. *J. Am. Chem. Soc. 77*, 2351–2361.

van Niel, C. B. 1931. On the morphology and physiology of the purple and green sulphur bacteria. *Arch. Mikrobiol. 3*, 1–18.

Pirson, A., C. Tichy, and G. Wilhelmi. 1952. Stoffwechsel und Mineralsalzernährung einzelliger Grünalgen. I. Vergleichende Untersuchungen an Mangelkulturen von *Ankistrodesmus*. *Planta 40*, 199–253.

Urey, H. C. 1952. *The Planets*. Yale Univ. Press, New Haven.

VIII. ON THE ORIGIN OF SELF-REPLICATING SYSTEMS

BY HAROLD F. BLUM[1]

A DISCUSSION of this subject may properly begin with an examination of essential properties of modern cells, as a starting point for extrapolation back toward the primitive systems from which those cells have arisen. If we proceed in this way we remain always aware that there are certain functional properties that had to emerge—whether simultaneously or in sequence—in order for systems to come into being from which the present living organisms could have evolved. The latter is an aspect of the problem of origins that seems sometimes to be lost sight of. It is really not enough to picture how certain chemical compounds could have been formed, nor how cyclic or autocatalytic reactions could have been set up in a primitive inorganic environment. What we really seek to know is how there came into being a particular kind of self-replicating system which contained those properties that have made possible the evolution of living systems to their present complexity; systems that could be subject to natural selection—I use the term carefully, in the sense of modern Darwinism. Looked at in this way, the problem of the origin of life and the problem of organic evolution are not separable; understanding the one depends on understanding the other.

I am going to distinguish, for purposes of discussion, three functional properties of self-replication in living organisms, which seem to me to constitute a basic minimum for describing this process. I shall call them (1) energetic, (2) kinetic, and (3) spatial properties; but while I use these terms in their physical sense, I am concerned here only with their application to living systems. In this biological sense these are not truly separable properties, but are functionally interdependent; and the more one considers them in detail the more he sees how entangled they really are—that any distinction such as I am making is only valid in an abstract analytic sense. Indeed, the inseparability of these basic functional properties has in itself strong implications with regard to Life's origin; and this becomes only the more apparent when we try to separate them.

Let us look at the energetics of replication in very general and elementary terms. In the living organism, parts have to be synthesized and

[1] National Cancer Institute, Public Health Service, Department of Health, Education, and Welfare, Bethesda, Maryland. Also, Department of Biology, Princeton University (present address).

put together in some form of pattern, say (by way of analogy) against a template. Then, if replication is to have any biological significance, the newly formed pattern has to be taken off the template. Ultimately, both replica and template have to be broken down into constituent parts in order that these may be used again; for one of the characteristics of living systems in general is that they use the same material over and over—if they did not, evolution would have ceased long, long ago. Now, somewhere in this cycle work has to be done, which means that free energy is expended. For example, if the parts assemble themselves on the template spontaneously, work has to be done to take the replica off; or, if the replica comes off the template of its own accord, work must be done to put the parts on in the first place. If we could break down the energy balance in detail we might find that both types of process are combined in replication in the living cell. But in any event, when the total is tallied up there has to be an over-all loss of free energy—otherwise, we should have a perpetual motion machine.

To do the work essential to replication the cell must make use of an available supply of compounds capable of yielding free energy. The means of getting this energy lies in the domain of other papers in this symposium, on which I will not encroach, but will concern myself with use of the energy once it has been captured. I will only say that Dr. Gaffron's idea that photosynthesis originated quite independently of the origin of replicating processes makes good sense to me.

One of the striking characteristics of cells is that they are able to release the potential energy of such compounds as, say glucose, in a series of steps, making it possible to tap off small quantities that may be used for doing work of one kind or another. By means of energetically coupled reactions, some of this free energy is used in essential syntheses that require the addition of free energy; that is, they represent endergonic steps that will not go spontaneously, as contrasted to exergonic steps which will go if left to themselves. Very often, as is well known, the transfer of free energy in these endergonic steps involves the exchange of "energy rich"[2] phosphate bonds, in the ATP-ADP cycle. In losing an energy rich phosphate group, adenosine triphosphate, ATP, becomes adenosine diphosphate, ADP. If another energy rich phosphate group is added from some other energy rich source, ADP is returned to ATP. By means of this cycle there is a repeated transfer of free energy in

[2] While it might be more correct to speak of energy rich group transfer, the term energy rich bond is in such common use in biochemistry that it need not introduce any confusion (e.g. Lipmann, 1955).

small amounts from one compound to another, and through this mediation it is possible to accomplish a variety of syntheses, including the work of transport, etc., necessary thereto. The amount of energy that can be transferred by means of the ATP-ADP reaction is relatively small, in the neighborhood of 10–15 kg cal/mole, and endergonic reaction steps which can be forwarded by such energy transfer are, of course, limited to such as require no greater quantity of free energy. There are other kinds of transferable energy rich bonds in living systems, notably the acyl bonds, whose transfer corresponds to about 16 kg cal/mole (Lipmann, 1955). But whatever endergonic processes go on in the cell are apparently limited to relatively small changes in free energy, and this would seem to lend a flexibility to energetic changes in these systems which would not exist if energy had to be used in larger spurts.

The energy changes in the cell might be diagrammed roughly as in Fig. 1, where the exergonic stepwise release of free energy from an en-

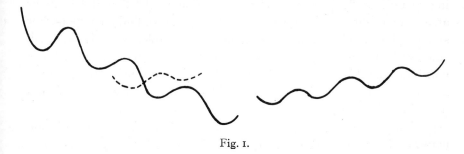

Fig. 1.

ergy rich source is represented by the descending line at the left. A single endergonic step is suggested by the dotted line which illustrates the necessary limitations on the size of that step. We may represent the building up of a compound such as a protein, about which I will have more to say presently, by a series of such small steps indicated by the ascending line on the right hand side of the diagram.

Now while such a scheme is described in the same terms and is consistent with the same thermodynamic principles as those that govern all chemical reactions, there is a difference in degree of complexity which distinguishes the living from the nonliving. Such a modulated cyclic process taking place in step-wise fashion is not, I think, paralleled in any nonliving system that we know today; nor for that matter, is it characteristic of the kind of systems that are generally speculated upon as existing on the surface of this planet prior to the advent of life. I think the distinction becomes more and more apparent as we examine into the other functional properties I have mentioned above. We all believe that living systems were at some time in the distant past derived from non-living ones; but we must ask ourselves how this transition in degree of complexity was accomplished. No doubt much of the detail that now characterizes living organisms was worked out in the long, slow process of natural selection; but I find difficulty in visualizing how some of the basic properties common to all living systems could have arisen in this way, because natural selection is based on the existence of self-replicating systems of a quite complex order. That a good deal of "chemical evolution" went on at an early date before the advent of living systems, seems incontestable; and this no doubt involved a kind of selection process. But I think one should be careful not to confuse this with Darwinian natural selection (see Blum, 1955).

In the present connection, let me illustrate the problem of transition by returning briefly to the ATP-ADP system, which is found participating so universally in energy transfer in living cells that it has to be assumed a common constituent of all. And when we find a compound or system playing ubiquitously such an essential role, we may well ask whether it was not a component of the first replicating systems. Or if not, how was this essential role carried out in those primitive systems, and how did it come later on to usurp this role so completely? Is it not likely in such a case that the system concerned was a product of chemical evolution prior to the advent of replicating systems, in which evolution natural selection did not participate?

But whatever its origin, the ATP-ADP reaction and all the kinds of group transfer essential to life processes today, require specific catalysts, and this brings us to a consideration of kinetics. Thermodynamic properties, specifically the free energy of reaction, tell us whether a reaction *can* go, but tell nothing about how fast it *will* go. In the diagram (Fig. 1) I have indicated each step, whether "uphill" or "downhill," occupying a depression, so that to move on to the next step a certain energy barrier has to be overcome. In terms of the theory of absolute reaction rates (e.g. Johnson, Eyring, and Polissar, 1954) this is the *free energy of activation*, and it is this which regulates, basically, the speed of a reaction. One way to overcome the energy barrier is to lower it, and this is essentially what happens when reactions are speeded up by catalysis. The rates of reactions in living systems are moderated by the catalytic activity of enzyme systems, usually specific for a given reaction; and while a catalyst cannot supply the free energy to an endergonic reaction, no endergonic step takes place in a living organism without the mediation of some enzyme system which determines its rate of achievement. This is true of the reactions involving the ATP-ADP system or any other transfer of energy rich groups.

The enzymes, which are a component of every catalytic system in living cells, are proteins—substances which are, hence, truly indispensable to replication as well as to virtually all the other dynamic processes that characterize living cells. Were such compounds present in the first replicating systems? If not, what played their indispensable role? And how was the transition accomplished from a "more primitive" form of catalysis to catalysis by enzymes? Before venturing any answer let us consider briefly some essential features of proteins.

These compounds are generally regarded as long polypeptides—polymers in which the monomers are amino acids. In this chain-like structure the amino acid monomers are joined together by peptide linkages, the formation of which may be represented by the reaction

$$\text{amino acid}_1 + \text{amino acid}_2 \rightleftarrows \text{dipeptide} + H_2O;$$
$$\Delta F \cong 2.75 \text{ kg cal/mole}[3]$$

This reaction entails the removal of H_2O; it is the opposite of hydrolysis. The free energy ΔF is positive, which means that the reaction would not go spontaneously, but would require that a small amount of free energy be added to make it go. The amount of free energy required to form a peptide linkage is, however, within the compass of a single energy rich phosphate bond, and it may be reasonably supposed that in the living cell the free energy for forming each peptide linkage is supplied by the transfer of a single energy rich bond (e.g. Lipmann, 1954; Chantrenne, 1953). It would be hard to devise a mechanism for supplying in a single step all the energy needed for forming peptide linkages in a protein molecule; but if this can be done a step at a time the problem is easy insofar as energetics is concerned.

There is more to protein structure than the simple polypeptide chain. The peptide linkages constitute the primary bonds; but secondary bonds, conspicuously hydrogen bonds, may hold the polypeptide chains together in helical form; and tertiary bonds, consisting chiefly of side linkages between amino acid groups, may hold the chains together in a variety of configurations which have much to do with the specificity of protein structure (e.g. Linderstrøm-Lang, 1952; Pauling and his group,

---

[3] This is a rough "average," values of ΔF ranging from about 2–4 kg cal/mole, for the cases measured. (See Linderstrøm-Lang, 1952, for a discussion of the thermodynamics of polypeptide formation.)

1951). Thus, when we think about the formation of proteins we have to think of the free energies concerned in forming all these bonds. The thermodynamics of this situation has not been completely elucidated; but it seems reasonably clear that the hydrolysis of proteins to their component amino acids is exergonic in dilute aqueous solutions, because it can be readily catalyzed by appropriate enzymes. Conversely, we may reason that the over-all formation of these compounds under such conditions must be endergonic. Thus, while protein formation may be accomplished in living cells which are equipped to do the necessary work, the spontaneous formation of these compounds in a non-living milieu presents energetic problems that need to be taken into account in considering origins.

In living organisms the transfer of energy rich bonds, which would seem to be involved in protein formation as well as many other syntheses, is catalyzed by enzyme systems, so the presence of proteins is essential. Can we presuppose that proteins, or at least long polypeptides, were present in the first replicating systems? For purposes of orientation it may be interesting to make a simple calculation of the probability of a long polypeptide being formed if this depended only on peptide linkages according to the reaction written above. The free energy assigned there corresponds to an equilibrium condition at which in a molal solution of amino acids, 1% of these (i.e. 10^{-2}) would be combined as dipeptides. At the same time it would be expected that 10^{-4} of the amino acids would be combined as tripeptides, 10^{-6} as 4-peptides, etc. The smallest protein contains about 80 peptide linkages, and the probability of correspondingly long polypeptides in the milieu we have postulated, would be about 10^{-160}. Even if enough nitrogen could be found,[4] we see that the chance formation of a single molecule of an 80-peptide under these conditions is extremely improbable.

But we should not, I think, use this sort of evidence to support a miraculous beginning,[5] or the idea that life began as a single highly improbable event.[6] While the above calculation indicates the improba-

---

[4] There are only about 3×10^{20} gram atoms of N in the earth today, almost all of which is in the atmosphere, and it is unlikely that there ever was much more of this element present since the planet became a discrete mass.

[5] A calculation of about this order was accepted by the late Lecomte du Nouy as evidencing the need for extraphysical explanation of life processes. (E.g. in *Human Destiny*, Brentano, New York, 1947.)

[6] The great length of time available is sometimes invoked to account for such an improbable event, and it may be worth pointing out with regard to the above calculation that time favors the approach to equilibrium. I have discussed elsewhere the difficulty of the extreme position that life originated as a single chance event, and of the opposite extreme that life as we know it must have arisen inevitably (Blum, 1955).

bility of the first proteins forming spontaneously in dilute aqueous solution, it is not difficult to imagine conditions under which the equilibrium might be more favorable to the formation of long polypeptides; and, moreover, the calculation is based on estimates that may exaggerate the difficulty.

It is seen from the reaction for the peptide linkage that the removal of water from the environment should favor the formation of these linkages, and the preparation of artificial polypeptides in the laboratory is usually carried out in nonaqueous media (e.g. Katchalski, 1951). It would not seem too implausible that the first polypeptides were formed under more or less anhydrous conditions. From the experiments of Miller (1953, 1955) we may believe that amino acids were formed in the earth's primitive atmosphere, and that they were carried down in water droplets to accumulate in the surface waters of the earth. We may imagine that with the beginning of orogenic changes in the earth's crust, pools of amino acid solution were formed, of greater or less extent. If the water evaporated from some of these pools, the formation of polypeptides should have been favored, and if with further topographic changes the residue was redissolved, there could have been present polypeptides in solution and suspension. Such polymerization might also have taken place in the nonaqueous phase of a coacervate,[7] or possibly there were other conditions in the then existing nonliving milieu which favored the formation of long polypeptides. Once formed, and lacking catalytic forwarding of their hydrolysis, the long polypeptides could conceivably have remained for a long time, and have been ready at hand to play a role in the first replicating systems.[8]

The objections to forming polypeptides in dilute aqueous solution need not apply to insoluble polypeptides, where the equilibrium may favor the formation of peptide linkages.[9] It is conceivable that insoluble polypeptides might act as catalysts, particularly if adsorbed on surfaces, although such a possibility does not seem to have been explored. It is also possible that small polypeptides, which would be more likely to form, would display catalytic activity if we studied them. It is customary to study enzyme action in solution, and here the catalytic efficiency de-

[7] A role of coacervates in the origin of life was suggested by Oparin (1938).

[8] The suggestion that the first polypeptides were formed under nonaqueous conditions, which I have presented elsewhere (1954, pp. 178c; 1955) is also favored by Fox (e.g. Fox and Middlebrook, 1954), who has been concerned with the origin of proteins as a result of thermal reactions at the earth's surface. He has found both amino acids and polypeptides to be formed under conditions of "dryness" and moderate heat.

[9] E.g. Tauber (1951a, b) has obtained insoluble, long, polypeptides enzymatically from protein free hydrolysates.

pends upon the intact molecule,[10] but we need not postulate that the catalysts in the first replicating systems were either as efficient or as specific as the enzymes in modern cells.

So the formation of polypeptide ancestors of the modern proteins before the advent of self-replicating systems seems not so improbable as at first appears, particularly if they did not have to form in a "dilute hot soup." The alternate hypothesis that the ancestral replicating systems were nonprotein seem to offer evolutionary difficulties that may be harder to surmount.

It becomes clear that the thermodynamic and kinetic aspects of the problem of origin of living systems cannot really be separated; and spatial pattern, which I want to discuss now, seems tangled up with both. The intimate relationship between biochemical activity and morphological structure in the cell becomes increasingly apparent as study presses forward at microscopic and submicroscopic levels; but there is another basic aspect of spatial pattern that concerns us particularly with regard to origins. This is the all important function of reproducing and handing on the essential pattern from one cell to the next, which is the basis of heredity. In order to preserve the character of the species the pattern must be handed on accurately from one cell to another. But in order for new species to evolve, the pattern must be subject to error, and moreover, this error must sometimes be itself replicable—this is the essence of mutation. So replicating systems which could give rise through evolution by natural selection to the complexity of existing species, must have been based on patterns which were at the same time essentially stable, and modifiable in minor respects.

Such stability of essential pattern, with at the same time the possibility of minor replicable error, is offered by such long polymers as the proteins, where the variety of monomers makes possible a tremendous variety of combinations and permutations, and where change of a single monomer need not greatly alter the pattern as a whole. The specific character of enzymes depends upon their molecular configuration, which has in some way to be replicated within the cell, and from cell to cell. But it becomes increasingly clear that the master patterns of inheritance of cells are associated with another type of polymer, the nucleic acids, particularly one type of these, desoxyribose nucleic acids or DNA. The monomers in these polymers are nucleotides, each of which is comprised

---

[10] A single quantum of ultraviolet light may inactivate an enzyme, and this could hardly do more than rupture a single bond, since a chain reaction is not involved (e.g. McLaren, 1949).

of a purine or pyrimidine base, a five carbon sugar and phosphate, connected as follows:

$$\text{base} - \text{sugar} - \text{phosphate}$$
$$\text{base} - \text{sugar} - \text{phosphate}$$
$$\text{base} - \text{sugar} - \text{phosphate}$$

The number of monomer types is smaller here than in the case of the proteins, but there are many monomer units in the nucleic acid polymer, and hence opportunity for a wide variety of combinations and permutations.

As in the case of the peptide linkage, the joining of the monomers of nucleic acid requires the removal of one molecule of H_2O per linkage. According to Watson and Crick the DNA polymer is coiled in a helix held together by hydrogen bonds, and as in the case of the proteins we are without direct information regarding the thermodynamics of formation of these bonds. But again it seems reasonable to assume that the formation of nucleic acids in the cell is in an over-all sense endergonic, their hydrolysis exergonic, since like the proteins these substances are subject to catalytic hydrolysis under mild conditions. It is also reasonable to suppose that the endergonic phase of the replication of these substances in living organisms takes place in small free energy steps, such as can be encompassed by transfer of single energy rich groups. And again it seems not unreasonable to think that, as I suggested for the polypeptides, the nucleic acids were originally formed prior to the advent of self-replicating systems; their formation being perhaps favored by dryness or other conditions tending to shift the equilibrium in that direction. It is interesting to go a little further and to speculate that polypeptides and nucleic acids were formed in the same puddle, or puddles, as these dried up; and that the two types of substances have been associated ever since. For these two polymers are inseparable constituents of all living cells; and evidence accumulates to indicate that nucleic acids take part in protein synthesis, while proteins serve as catalysts in the synthesis of nucleic acids.

Recently, Grunberg-Manago, Ortiz, and Ochoa (1955) have been able by means of an enzyme to polymerize diphosphates (pyrophosphates) of nucleotides to obtain nucleic-acid-like chains, with the release of orthophosphate to the medium. The reaction is quite reversible, with the equilibrium favoring the formation of the polymer. Whatever the relationship of this reaction to the formation of nucleic acids in vivo, it would seem to offer a kind of reaction that might have gone on in the primitive nonliving milieu, granted appropriate catalysis.

It may be pointed out that in both the case of the polypeptides and nucleic acids, the polymerization is at least fairly reversible, so that it may be readily shifted by changes in the milieu. Thus the formation of these substances in the nonliving environment becomes reasonably likely, granted the existence of the appropriate monomers. But in the replicating system it would seem advantageous if the monomers could be put together one or a few at a time, so that the environment as a whole would not have to be changed. This seems to be what occurs in the living system by virtue of its ability to transfer energy locally in small quantities by means of energy rich group exchange. This step by step synthesis, without the attendant instability that would result from wholesale changes in equilibrium, would seem to be an essential characteristic of successful replication, and one which distinguishes the living from the nonliving. Thus, while we may postulate general shifts of equilibrium with regard to the formation of the materials for the first replicating systems, the successful emergence of those systems themselves would seem to require some means for energy transfer in small steps.

Taken for granted above is the pre-existence of nucleotides, and this may offer some difficulties. Not only are these substances the monomers of nucleic acids; but ATP, which would seem the most likely candidate for the essential role of free energy transfer in the first replicating systems, also has a nucleotide structure, although it has not up to now been found present in nucleic acids. Now the origin of nucleotides offers a particular problem because they contain phosphorus, and this is quite a rare element on the surface of the earth. We can, following Urey (1952a,b) and Miller (1953; 1955), picture the formation of amino acids in the primitive reducing atmosphere that surrounded our earth, where the essential component element, nitrogen, was present in high proportion; but it is more difficult to imagine phosphorus to have been there, so the nucleotides, just as indispensable components of living systems as the amino acids, probably had their origin in some quite different way. Compelling and important as Miller's experiments are, let us not think that they have solved the problem of the origin of life, as I am sure their author would be the first to recognize.

Gulick (1955) has pointed out that under the anaerobic conditions pertaining on the primitive earth (Oparin, 1938; Urey, 1952), phosphorus could have been present in soluble form, but concentration of this element would seem to have required erosion of the earth's surface. This assumes the commencement of orogeny, and suggests again that life arose in a local concentrated solution rather than a dilute one, since it has to be assumed that a long time was required to reach the present concentration of salts in the ocean as a whole. Gulick also suggests a pathway for the formation of energy rich phosphate bonds, but again one has to wonder how the reactions were accomplished in the absence of enzymes. Nevertheless, one may suspect that the origin of nucleotides may turn out to be less obscure than currently appears. It seems probable that more exploration in terms of chemistry and thermodynamics regarding the compounds that might have been formed under conditions simulating those that could have pertained at the time life came into being, may lead to a much better view of the matter. I suspect that a good many problems might disappear in this way, provided proper attention is paid to biological and geological requirements. Miller's experiments provide a brilliant start in this direction.

I have stressed that the presence of certain substances playing the same indispensable roles in all modern cells indicates that these substances, or close analogues, were present in the original replicating systems from which these cells have evolved. I have named specifically, ADP, proteins, and nucleic acids, but more could no doubt be included. The reader may ask whether the original replicating systems could not have been of some different, more primitive type, from which the ones we know evolved later on. I will have to answer first with a question. Would we not expect that if this were the case, some vestige would be left of these more primitive systems? Would we not expect that among the vast array of existing species, some organisms might still be using the primitive way of replicating? When Bernal (1951) suggests that catalysis on the surface of clay particles was a precursor of the catalysis essential to modern living organisms, Pringle (1954) asks why in that case aluminum is not a prominent component of modern cells, instead of a very rare and inconstant element. Bernal (1954) acknowledges the weight of Pringle's argument, which is one that I think might be expanded with profit to our thinking about origins in general. I might point out that having polypeptides formed under favorable conditions, could remove the need of catalysis by clay or any other inorganic system.

Natural selection is often invoked in such cases to get across the gulf separating the nonliving from the living; and here we meet a fundamental difficulty. Natural selection in the modern Darwinian sense—the only way in which the term should be used, I think—implies a replicating system that can hand on a pattern with only occasional or rare modification. Long polymers, such as proteins or nucleic acids are fitted, by their large size and repetitive structure, to take care of this kind of function, and the constitution of the patterns concerned in gene inheritance seems to be narrowing down to one type of the latter, DNA (e.g. Hotchkiss, 1955). Now it seems to me that if there is a plausible scheme by which polypeptides and nucleic acids could have been formed under favorable conditions, so that thermodynamic principles do not have to be violated, and so that their evolution from pre-protein, pre-nucleic-acid replicating systems does not have to be explained, we ought to give that scheme serious consideration. The whole thing seems easier that way.

I should point out that in speaking of primitive replicating systems I do not necessarily mean that the first of these were unitary, in the sense of being cellular. The patterns might, for instance, have been provided by nucleic acid molecules floating more or less free in a medium containing polypeptide catalysts, and molecules having energy rich phosphate groups such as ATP. Horowitz (1945), in formulating his very astute hypothesis about biochemical evolution, had, I suspect, something like this in mind.

But now I should say something more about spatial pattern with regard to origins. Obviously, I am speaking of molecular pattern, which must be the basis of most, if not all, of the microscopic pattern of cells. We are accustomed to think of the replication of parts as taking place, in some way, against a template. Although the nature of the template is usually not defined it seems obvious that it must have its basis in molecular configuration, and that it is not strictly a spatial pattern but a spatio-energetic one. Now a convincing feature of the template hypothesis is to be found in the existence in so many of the components of living cells, of one of two possible optical isomers to the exclusion of the other. For example, in nucleic acids the sugar component is invariably a D-sugar: D-ribose or D-deoxyribose. In proteins all the amino acids correspond to the L-form of lactic acid, except glycine which is not optically active (see Dunn, 1943). We know that there is no thermodynamic difference between mirror image isomers—both require the same amount of information to describe, and so have the same entropy. Being thermodynamically identical they cannot be separated by direct chemical means. Yet the living organism does separate them, and the only reasonable explanation is that it does so by replication against some spatial pattern—some template already containing optically polarized elements. In this way, the formation of a long polypeptide composed of only one optical antipodal type is just as probable as not in a system that can do the work necessary for assembly; whereas if such a polypeptide had to be formed by chance unscrambling from the expected fifty-fifty mixture it would be highly improbable.[11]

Now it can be seen where we are heading—straight into the problem of how the original template got formed. This is seemingly one of the most difficult questions regarding the origin of those replicating systems from which modern organisms could have evolved; and one that may need quite different solution from that of replication of patterns against a template already in existence.

There are a number of ideas current regarding the origin of the optical isomerism of components of living organisms. One of these, which I think was first proposed by Byk in 1904, invokes the action of polarized light, which he supposed to have been prevalent at the primitive earth's surface. Subsequently, a few cases have been described in which circularly polarized light has been found to forward the formation on one optical antipode relative to its mirror image (e.g. Mitchell and Dawson, 1944). But the reaction must be very favorably chosen if a high efficiency of separation is to result, as would seem necessary to account for the complete absence of one antipode from living systems in such cases as I have mentioned. It is also necessary to account in some way for the prevalence of circularly polarized light in the primitive surroundings.

Another suggestion is that all living things arose from a single one that happened to have a particular optical configuration; but on the basis of chance alone this seems to invoke a miracle several times repeated. Another is that the first components of living systems were formed against some pattern, say quartz crystals, that were of only one optical antipode, but this again poses the question how the particular antipode was chosen for the "template" crystals.

Still another suggestion, which I have tried myself to formulate (1954, pp. 178A–F), but not very satisfactorily, invokes natural selection. Let us illustrate with the case of proteins. It is conceivable that polypeptides built from only the L-form of amino acids would be more efficient catalysts than polypeptides composed from a mixture of D- and L-

[11] Suppose we start with an L-amino acid in a mixture of one-half L-, one-half D-forms. Other things being equal, the chances of joining with a second L-amino acid to form a dipeptide is one-half or 2^{-1}. The chances of this L-dipeptide joining on another L-monomer is again one-half, so the chances of forming a tripeptide of all L-forms is 2^{-2}, etc. On this basis the chance of forming an 80-L-polypeptide is 2^{-80} or 10^{-24}.

forms; and that they might be better structural units as well. This could provide a "handle" for natural selection, but there is no reason why the opposite, D-form should not be just as efficient as the L-. One can only suggest that the selection of the "left-handed" form depended upon some unrelated aspect of the replicating system—some independent handle for natural selection, having no relationship to the optical isomerism. But on this basis it is still difficult to eliminate the D-form because its occurrence is just as probable as the L-, and if it had not been eliminated once and for all early in evolutionary history we might expect to find it popping up somewhere among the great variety of existing species. This too seems to be far from a satisfactory hypothesis.

I do not want to pursue the discussion of optical isomerism further, but only to demonstrate with it the essential difference between the replication of forms against an existing pattern, and the origin of the pattern or template itself. The former, including the replication of the template, is a process that goes on continually throughout the living world, whereas the latter has not, to the best of our knowledge, been repeated for a very long time. Some of the experiments that are going on in laboratories at present, where viruses broken down into lesser parts reconstitute themselves (e.g. Fraenkel-Conrat and Williams, 1955), may eventually show us a good deal about how replication against an existing pattern takes place; but they seem likely to show us less about how the patterns originated in the first place. I cannot help drawing analogy with the taking apart and putting together again of a watch (and having it run) which, while difficult enough for most of us, presents considerably fewer problems than the original design of watches. The latter, like organic evolution, has its historical aspect. I use the word "design" metaphorically, of course, with no intention of introducing vitalistic or extraphysical concepts.

The distinction between replication of pattern and origin of that pattern may seem a subtle one at first, but grows more compelling as it is pondered. Twenty-five years ago a great pioneer in the area between physical chemistry and biology, Sir William B. Hardy, drew the distinction very clearly with regard to optical isomerism. He did this in an essay entitled *To Remind* (1934), presumably because he felt that the problem had been lost sight of. I think that today, in spite of the great advances in our knowledge, we may well be mindful of the questions he posed concerning this and other aspects of the problem of origins.

BIBLIOGRAPHY

Bernal, J. D. 1949. The physical basis of life. *Proc. Physical. Soc. 62*, 537–558. Reprinted in book form, Routledge and Kegan Paul, London.

Bernal, J. D. 1954. Origin of Life. In *New Biology*, pp. 28–40. Penguin Books, London.

Blum, H. F. 1954. *Time's Arrow and Evolution*, 2nd ed. Princeton Univ. Press, Princeton.

Blum, H. F. 1955. Perspectives in evolution. *Am. Sci. 43*, 595–610.

Blum, H. F. Behind natural selection. (In course of preparation.)

Byk, A. 1904. Zur Frage der Spaltbarkeit von Razemverbindungen durch zirkular-polarisiertes Licht, ein Beitrage zur primären Entstehung optisch-aktiver Substanz. *Ztschr. Phys. Chem. 49*, 641–687.

Chantrenne, H. 1953. Problems of protein synthesis. *Symp. Soc. Gen. Microbiol. 2*, 1–20.

Dunn, M. S. 1943. The constitution and synthesis of the amino acids. In *Addendum to the Chemistry of Amino Acids and Proteins*, ed. by C. L. A. Schmidt, Chapter II, pp. 1035–1043. Charles C. Thomas, Springfield.

Fox, S., and M. Middlebrook. 1954. Anhydrocopolymerization of amino acids under the influence of hypothetically primitive conditions. *Federation Proc. 13*, 211.

Fraenkel-Conrat, H., and R. C. Williams. 1955. Reconstitution of active tobacco mosaic virus from its inactive protein and nucleic acid components. *Proc. Nat. Acad. Sc. 41*, 690–698.

Grunberg-Manago, M., P. J. Ortiz, and S. Ochoa. 1955. Enzymatic synthesis of nucleic acid-like polynucleotides. *Science 122*, 907–910.

Gulick, A. 1955. Phosphorus as a factor in the origin of life. *Am. Sci. 43*, 479–489.

Hardy, W. B. 1934. *To Remind, A Biological Essay.* (The Abraham Flexner Lecture 1931.) Williams and Wilkins, Baltimore. And 1936 in *Collected Papers*, Cambridge Univ. Press, Cambridge.

Horowitz, N. H. 1945. On the evolution of biochemical synthesis. *Proc. Nat. Acad. Sci. 31*, 153–157.

Hotchkiss, R. D. 1955. The biological role of deoxypentose nucleic acids. In *The Nucleic Acids*, Vol. II. Ed. by E. Chargaff and J. N. Davidson, Chapter 27, pp. 435–473. Academic Press, New York.

Johnson, F. H., H. Eyring, and M. J. Polissar. 1954. *The Kinetic Basis of Molecular Biology*. John Wiley and Sons, New York.

Katchalski, E. 1951. Poly-α-amino acids. *Advances in Protein Chemistry 6*, 123–185.

Linderstrøm-Lang, K. V. 1952. *Proteins and Enzymes*. Stanford Univ. Press, Stanford, California.

Lipmann, F. 1954. In *The Mechanism of Enzyme Action*, ed. by W. D. McElroy and H. B. Glass. Johns Hopkins Press, Baltimore.

Lipmann, F. 1955. Coenzyme A and Biosynthesis. *Am. Sci. 43*, 37–47.

McLaren, A. D. 1949. Photochemistry of enzymes, proteins and viruses. *Advances in Enzymology 9*, 75–170.

Miller, S. L. 1953. A production of amino acids under possible primitive conditions. *Science 117*, 528–529.

Miller, S. L. 1955. Production of some organic compounds under possible primitive earth conditions. *J. Am. Chem. Soc. 77*, 2351–2361.

Mitchell, S., and I. M. Dawson. 1944. The asymmetric photolysis of β-chloro-β-nitroso-αδ-diphenylbutane with circularly polarised light. *J. Chem. Soc.*, pp. 452–454.

Oparin, A. I. 1938. *The Origin of Life*. Macmillan, New York. Reprinted, 1953, Dover Publications, New York.

Pauling, L., R. B. Cory, and H. R. Bronson. 1951. Atomic coordinates and structure factors for two helical configurations of polypeptide chains. *Proc. Nat. Acad. Sci. 37*, 205–211 (and later articles in the same journal).

Pringle, J. W. S. 1954. The evolution of living matter. In *New Biology*. Penguin Books, London.

Tauber, H. 1951a. Synthesis of protein-like substances by chymotrypsin from dilute peptic digests and their electrophoretic patterns. *J. Am. Chem. Soc. 73*, 4965–4966.

Tauber, H. 1951b. Synthesis of protein-like substances by chymotrypsin. *J. Am. Chem. Soc. 73*, 1288–1299.

Urey, H. C. 1952a. *The Planets*. Yale Univ. Press, New Haven.

Urey, H. C. 1952b. On the early chemical history of the earth and the origin of life. *Proc. Nat. Acad. Sci. 38*, 351–363.

IX. SOME ENERGY TRANSDUCTION PROBLEMS IN PHOTOSYNTHESIS

BY BERNARD L. STREHLER[1]

I. INTRODUCTION

PHOTOSYNTHESIS, the process whereby green plants use light energy to convert water and carbon dioxide into stored organic material and atmospheric O_2, is the ultimate source of food for all animal life. While by some standards photosynthesis is a most typical biological process and can be considered as a formal reversal of respiration, in one way this process is distinct from all other biological processes: the first step of photosynthesis involves the conversion of light energy into chemical potential energy.

It is unlikely that a complete understanding of the mechanism of photosynthesis will enable us to improve on a series of reactions which have been several billion years in evolving. Unlimited foodstuffs are not a direct or foreseeable consequence of photosynthesis research, newspaper accounts to the contrary.

It is likely, however, that the understanding of the many mechanisms operative in this complicated process will have a more or less direct application to problems in non-photosynthetic organisms. Particularly is this the case with the mechanisms of energy transduction which are at the center of all biological processes. These are still the main unknown features of the sequence of reactions in photosynthesis.

Viewed from the standpoint of energy transductions, photosynthesis can be divided into three consecutive phases. Phase I consists of the transformation of radiant energy into excitation energy in a specific

[1] Department of Biochemistry, Fels Fund, University of Chicago. Present address: Gerontology Branch, National Heart Institute, National Institutes of Health, Baltimore City Hospitals.

The author wishes to thank the following persons for their helpful comments during the preparation of this manuscript: Dr. James Franck, Dr. Hans Gaffron, Dr. Wm. McElroy, Dr. John Platt, Dr. Norman Bishop, Dr. Theodore Reiff, and my wife, Theodora Strehler.

This work was supported in part by a grant from the U.S. Atomic Energy Commission and part of it was undertaken while the author was a visiting investigator at the Carnegie Institute of Washington Plant Biology Laboratory, Palo Alto, California. Dr. C. S. French and Dr. James H. C. Smith were of great assistance in various phases of the work, both by their consultation and through their kind permission to use specialized equipment they had constructed.

Thanks are also due to Mr. John Hanacek, Mr. L. R. Kruger, and Mr. Charles Soderquist for their assistance in the construction of apparatus.

molecule. Phase II consists of the conversion of this excitation energy into some "primitive" form of reducing and oxidizing equivalents which are then in turn converted into redox couples capable of interacting with fixed carbon dioxide and of releasing oxygen. Phase III involves the carbon chemistry of the process, and consists of (a) the conversion of the energy stored in unstable redox couples into more stable covalent bonds in carbohydrates which can be stored and (b) a cycle for continually supplying molecules to which carbon dioxide can be added.

Most workers in photosynthesis would agree that:

1. Light quanta are captured by one of a variety of pigments, e.g. chlorophyll a and b, carotenoids, phycobilins. These captured quanta are then transferred to chlorophyll a by a process of sensitized fluorescence.

2. The excitation energy in chlorophyll a is used to "split" water into a precursor to reducing material and a precursor of molecular oxygen.

3. The reducing material is used to reduce "fixed" CO_2 to carbohydrate.

4. The primary oxidant eventually gives rise to O_2.

5. The process is propagated through a regenerative chemical cycle which produces "CO_2 acceptors" from part of the "fixed and reduced CO_2" (see Fig. 1). (Bassham et al. 1954.)

6. High energy phosphorus compounds are formed under the influence of light and are then in turn consumed to drive various reactions in photosynthesis.

These statements reveal very little about the physical and chemical mechanisms in photosynthesis (except for items 1 and 5). The present discussion will deal with some recent experimental findings and interpretations directed toward the questions:

How is the energy in excited chlorophyll transformed into a primary reducing and oxidizing source?

What kind of molecules or structures are these primary oxidants and reductants and what are their redox potentials?

How do the products of several single quanta interact and collaborate?

How is reducing potential transported from the site of formation to the site of utilization?

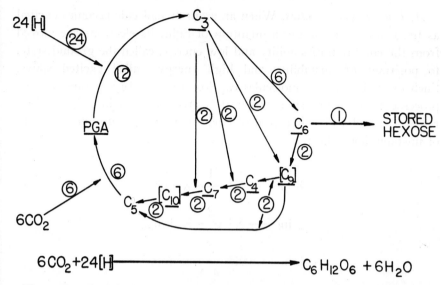

Fig. 1. The Carbon Dioxide Acceptor Cycle in photosynthesis. The circled figures indicate the number of units used per complete cycle. The underlined figures represent sugars or sugar phosphates of the chain length indicated. The bracketed figures represent hypothetical intermediates. Simplified from Calvin.

How is ATP formed?

What are the hydrogen carriers linking photochemistry to CO_2 reduction?

How is the primary oxidant transformed into molecular oxygen?

These questions generally lie in the area of the process linking the two better understood portions of photosynthesis: the physical act of light absorption and the chemical reactions which terminate the process. In other words, we do not know much about the links between the pure physics and the pure biochemistry of photosynthesis.

II. PHOTOCHEMICAL REACTIONS IN PHOTOSYNTHESIS

Early reactions in photosynthesis have most frequently been measured through the application of physical-optical techniques such as fluorometry, emission spectroscopy, and absorption spectroscopy. Photometric procedures are eminently suited to this kind of problem because they are rapid and sensitive, characteristics required of measuring methods designed to deal with chemical changes induced by light which are frequently transient and of small magnitude.

Unfortunately, results obtained through such physical methods are frequently difficult to interpret.

A. Fluorescence studies. When an atom or molecule becomes excited as the result of absorbing a quantum of light, its electrons are moved from the condition of stability and low energy, called the ground state, to positions of instability and high energy, called excited states. Such excited atoms or molecules can lose their energy by one of three processes: (1) conversion into heat, (2) conversion into light (fluorescence), and (3) conversion of excitation energy into another form of unstable bond, i.e.:

$$X + h\nu \longrightarrow X^* \xrightarrow{\;1\;} X + heat$$

$$X + h\nu \longrightarrow X^* \xrightarrow{\;2\;} X + light$$

$$X + h\nu \longrightarrow X^* \xrightarrow[+A]{\;3\;} X + A'$$

Experimentally, we can measure process 1 with considerable ease. Therefore, if we know the rate of formation of X^*, we can deduce the value of (1)+(3) by measuring (2). In practice, we are generally interested in the rate of (3) (since this includes chemical reactions), but unless we can be certain that (1) is constant, we cannot be sure of the value of (3). Unfortunately, there are a great many factors which can affect the rate of (1) in a living system so that there is no unequivocal interpretation of fluorescence changes during the course of photosynthesis. For example: It would be convenient to assume that the changes in fluorescence which occur when green plants are illuminated reflect the inverse of the rate of the chemical reactions storing light energy (see Fig. 3). Since it has been shown that molecular oxygen will quench the fluorescence of chlorphyll in solution by facilitating reactions of type (1) and since O_2 is a product of photosynthesis, it is uncertain whether the changes in fluorescence depicted in the figure reflect changes in oxygen concentration, changes in the concentration of energy acceptor for process (3), both or neither.

Despite these difficulties, a number of fundamental facts about photosynthesis have been deduced from chlorophyll fluorescence studies. W. A. Arnold (personal communication) has obtained evidence suggesting that light energy is passed about within the chloroplast from one chlorophyll molecule to another. He obtained this evidence through the measurement of the depolarization of the fluorescent light emitted by chlorophyll in the plant or in solution when it is exposed to polarized exciting light.

The polarization of fluorescence of a dye in solution depends on:

1. The viscosity of the solvent
2. The mass, size, and shape of the solute
3. The temperature
4. The lifetime of the excited state
5. The angular relationship between the absorbing and emitting optical axes of the molecule

In a solvent of given viscosity(v) and temperature(t), a given molecular species will rotate at a rate(r). This molecule will absorb polarized light of appropriate wave length when its optical axis and the axis of polarization coincide, but will not absorb when they are perpendicular. If it emits fluorescent light before it rotates, the fluorescent light will be polarized; conversely, if it rotates before it fluoresces, the emitted light will be depolarized.

In viscous solvents the rate of rotation of chlorophyll is so reduced that it fluoresces before it has an opportunity to turn appreciably on its axis. Therefore the fluorescence it emits when excited with polarized light is highly polarized.

In chloroplasts, where the individual molecules are part of a larger organized structure, one would expect a considerable restriction on rotational motion. However, the fluorescence emitted is highly depolarized. Arnold has interpreted this to mean that the light quanta are passed about from one chlorophyll molecule to another before a fluorescent quantum is emitted. It is postulated that it is the movement of the excitation energy from one chlorophyll molecule to another whose optical axis is somewhat inclined to the first, which results in fluorescence depolarization in the chloroplast.

Another application of fluorescence to the study of early steps in photosynthesis has been the finding by Arnold and Oppenheimer (1950) and others that light captured by fluorescent pigments absorbing further in the blue than chlorophyll a (such as phycocyanin) nevertheless causes the fluorescence of chlorophyll a. Further, they showed that this transfer of energy could only be explained as sensitized fluorescence. This observation means that regardless of which pigment system absorbs the light, it eventually results in the excitation of chlorophyll a and thus in the same photochemical reaction.

Sensitized fluorescence, which is the basis for both of the above, is the process whereby one excited dyestuff molecule transfers its energy to another molecule absorbing in the region of the first molecule's fluorescence emission. This reaction takes place with a much higher probability than would appear possible if one considered only the optical cross section.

What actually appears to happen is that the first molecule induces resonant oscillations in the electrons of the second molecule. This only occurs efficiently when the molecules are less than a few wave lengths apart, i.e. when the concentration of one of them is about 10^{-3} to 10^{-2} mole or more.

The above are two representative ways in which fluorescence has been applied to the study of the very early part of photosynthesis. Other findings of equal interest, but of less certain applicability to the intact photosynthesizing cell, have been made in the course of studies on chlorophyll photochemistry and fluorescence in vitro.

B. Absorption spectrum changes in photosynthesis. A second area in which optical methods have been applied successfully to the study of photosynthesis is absorption spectroscopy. Suppose that any one of the intermediates in photosynthesis has two chemical forms that possess different absorption spectra. Assume that one of these forms is increased during photosynthesis. It should then be possible to follow the rate of the reaction in which the substance participates, by measuring changes in the color of the photosynthetic tissues in the light and in the dark.

Such studies have been undertaken on chlorophyll solutions, on chloroplasts, and on intact photosynthesizing plants.

1. *In vitro studies.*

Of greatest potential significance is the finding by Livingston and Ryan (1953) and Linschitz and Rennert (1952) that chlorophyll *a* is converted (in organic solution) by light into a metastable form which possesses an altered absorption spectrum. This compound resembles in its spectral characteristics the compound formed when chlorophyll is illuminated in pyridine solution in the presence of ascorbic acid (Krasnovsky, 1948).

It has been suggested on numerous occasions that such a metastable form is involved in photosynthesis as the photo-produced intermediate possessing a sufficiently long lifetime to react in the slower enzymatic reactions following. Unfortunately all attempts to identify a similar intermediate in illuminated intact plants have been unsuccessful, and we therefore cannot assign any biological significance, at present, to these interesting findings.

2. *In vivo studies.*

Following Duysen's discovery (1954) that spectral changes occur in green plants during photosynthesis, a number of workers have studied one or another phase of this problem (Lundegardh, 1954; Witt, 1955; Coleman et al., 1956; Strehler and Lynch, 1956; Chance and Strehler, unpublished).

During illumination there is an increase in absorption at 515 mμ and a decrease in absorption at about 475 mμ (see Fig. 2; Duysens, 1954).

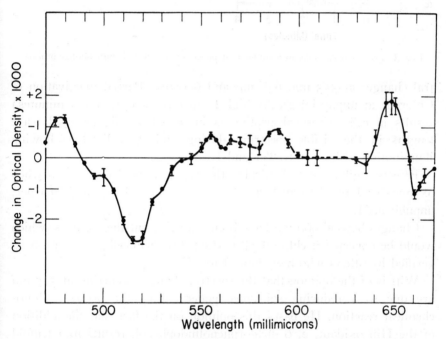

Fig. 2. Difference spectrum of Chlorella in light and dark (after Strehler and Lynch, in press).

With a single flash of light this effect is evident within 10^{-5} seconds and decays in about 0.01 seconds (Witt).

The effect shows an overshoot similar to that observed for other processes (e.g. fluorescence, luminescence, formation of ATP (see Fig. 3). Immediately following illumination there is a reversal of the above changes and then a slow return to the dark value. There are also spec-

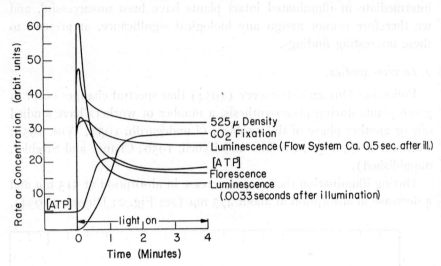

Fig. 3. Transient changes in a number of processes connected with photosynthesis.

tral changes at 555 mμ, 648 mμ and 660 mμ. Partial inactivation of Chlorella at unphysiologically high temperatures (51° for 4 minutes) results in cells whose absorption is increased at all measured wave lengths in the visible spectrum during and immediately following illumination (Fig. 4; Strehler and Lynch).

Effects similar if not identical to the 515-475 mμ changes occur when oxygen is admitted to an anaerobic suspension (Chance and Strehler, unpublished).

Changes beyond 660 mμ have been reported, but these effects, which would be expected if chlorophyll participates chemically, have not been verified by others who have looked for them.

Witt is of the opinion that the spectral changes occurring at 515 mμ are due to a reducing substance generated by the primary photochemical reaction. He bases this opinion on the fact that the addition of the Hill oxidant, 2, 6 dichlorphenolindophenol, results in a tenfold increase in the rate of disappearance of the 515 mμ band and because the dyestuff is concurrently reduced.

Strehler and Lynch, on less direct grounds, came to a similar conclusion. The recent findings of Chance and Strehler, that O_2 will substitute for illumination, however, make these conclusions less attractive.

The former authors' suggestion that the major spectral changes are due to a transformation of a flavin or carotenoid pigment is consistent with Chance and Sager's finding (personal communication) that a

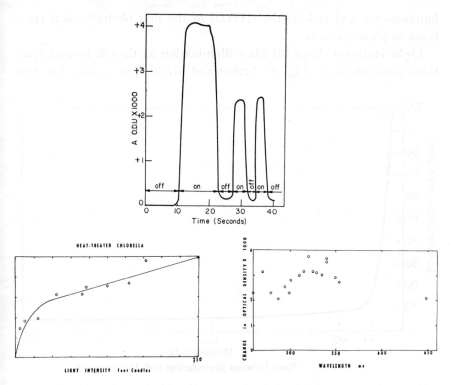

Fig. 4. The effect of heat treatment (51°C for 4 minutes) on the response of Chlorella to illumination. Flow system, measurements made at 515 mμ and 25°C. Top: Time course of transmission change; note the absence of induction effect. Right: The effect at different wave length settings of the monochromator; note the lack of sharp maxima. Left: Dependence of absorption changes on illuminating intensity; note approach to saturation.

mutant strain of Chlamydomonas which possesses a very low carotenoid content does not show the spectral change at 515 mμ. Whether this means that carotenoids are involved only in side reactions, secondary to the main reactions in photosynthesis, or whether it indicates that the diminished supply of carotenoid present in the mutant is used more effectively, cannot now be decided.

C. *Studies on photosynthetic luminescence.* A third aspect of the photochemistry of photosynthesis which has received attention in recent years is the phenomenon of photosynthetic luminescence discovered in green plants by Strehler and Arnold (1951). While attempting to measure ATP changes induced by the illumination of chloroplasts, they found that all photosynthetic tissues emit a dim luminescence after they are illuminated. The luminescence is in all likelihood a chemiluminescence and reflects the reversal of the first photochemical reactions in photosynthesis.

Light emission drops off after illumination as though several reactions were involved (Fig. 5; Arthur and Strehler, in press). The first

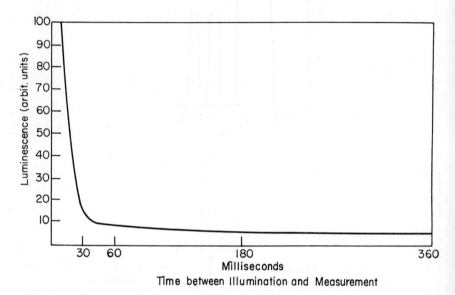

Fig. 5. Decay curve of Chlorella luminescence (from Arthur and Strehler, in press).

reaction decays in about 0.01 seconds while the slower one disappears with a half life of seconds. Arnold has detected some luminescence after more than 3 hours in the dark. The emitting molecule is chlorophyll a and the emission spectrum is experimentally indistinguishable from the fluorescent light emitted by chlorophyll a in vivo (Arnold and Davidson, 1954).

In contrast to fluorescence, luminescence saturates at high light intensities (see Fig. 6). The saturation curve for luminescence against incident light intensity is similar to that for the spectral changes at 515 mμ and 648 mμ (see Fig. 7; Strehler and Lynch, in press).

It has recently been found that dried chloroplast films conduct electricity when they are illuminated. Such preparations exhibit a dim luminescence which decays parallel to the conductivity after illumination (W. A. Arnold, personal communication). The nature of the chemical reaction responsible for luminescence is unknown (Bruegger, 1954), but it has been shown that chlorophyll a in solution chemilumi-

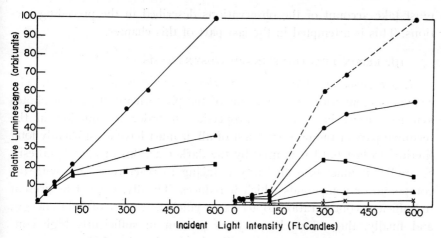

Fig. 6. Luminescence versus illuminating intensity for chloroplasts (left) and Chlorella (right). Time elapsing between flash and measurement (reading from bottom): 360 milliseconds, 180 ms, 3.3 ms (left) ; 360 ms, 60 ms, 30 ms, 6.6 ms, 3.3 ms (right). Chloroplasts suspended in 0.35 mole NaCl, 0.01 mole KCl. Temperature = 25°C (from Arthur and Strehler, in press).

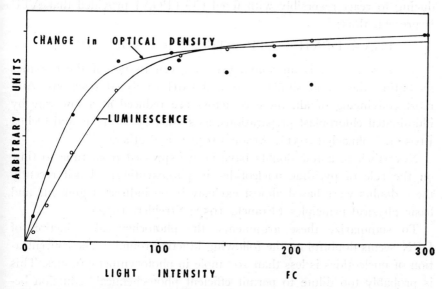

Fig. 7. A comparison of photosynthetic luminescence and change in optical density at 525 mμ in Chlorella at various intensities of photosynthesis-promoting light. Both effects were measured in a flow system, about 1 second after the cells were removed from the light (from Strehler and Lynch, 1956).

nesces in the presence of oxygen and aldehydes (Ferrari, et al., in press).

Any theory of the early photochemistry of photosynthesis obviously must take account of the observations described in the preceding sections. This is attempted in the last part of this chapter.

III. REDOX REACTIONS IN PHOTOSYNTHESIS

A. Evidence on the nature of the primary reducing agent. The reducing agent formed as a result of the photochemical part of photosynthesis must have several properties in order to function in the terminal part of the process. First of all, it must have a sufficiently long lifetime to last until it is used by the dark reactions of photosynthesis. Secondly, it must be sufficiently reducing to react reversibly with the compound or compounds which it reduces. Thirdly, it must be present in sufficient concentration to act as an efficient trapper of light energy; and finally, there must be enzymes present in sufficiently high concentrations to handle its dark utilization at the observed rates.

The pyridine nucleotides have received repeated attention from biochemists as likely candidates in the intermediate transfer of reduction energy, primarily because the pyridine nucleotides are sufficiently reducing to react reversibly with fixed CO_2(PGA) provided that ATP is present, thus:

$$PGA + DPNH + ATP \rightleftarrows Triose—P + ADP + DPN^+$$

Moreover, there is an abundance in green tissues of the enzymes mediating these and similar CoI and CoII catalysed reactions. And most convincing of all, these cofactors are reduced at a low rate by illuminated chloroplast preparations, as shown by Vishniac and Ochoa (1952), Tolmach (1951), Arnon (1951), and others.

Nevertheless, a few doubts have been expressed from time to time on the role of pyridine nucleotides in photosynthesis. Until recently these doubts were based almost exclusively on indirect arguments and basic physical principles (Franck, 1953; Strehler, 1952).

To summarize these arguments, the photochemical reduction of DPN seems doubtful on the following kinetic grounds: The concentration of nucleotides is less than 10^{-4} mole in photosynthetic tissues. This is probably too dilute to permit efficient photochemical reduction because the excitation energy will be dissipated before a collision takes place between the coenzyme and the excited dye molecule.

The following objections have been raised on thermodynamic grounds. The energy in the excited chlorophyll molecule is about 40 kcal/mole. It takes about 51 kcal to transfer 2 hydrogen atoms from water to DPN^+ (about 25.5 kcal/H). Therefore, if DPNH is to be formed by photochemical reduction of DPN^+, no more than 14.5 kcal of the initial 40 kcal can be wasted.

But any photochemical energy storage process is limited by the following fact. If the energy is stored by some reaction $X \longrightarrow Y$, the activation barrier for the reversal of the process, i.e. $Y \longrightarrow X$, must be greater than the activation barrier for reactions using X (i.e. $X \longrightarrow Z$, $X \longrightarrow A$, etc.) by an amount sufficient to prevent $Y \longrightarrow X$ from being an important reaction. Since there is practically zero probability of finding a large excess of thermal energy in the reactant simultaneously with the photochemical reaction, it follows that the barrier to the reverse reaction must arise through the dissipation of some of the energy in the excited molecule. Any energy used to stabilize the reactants is not available to do chemical work; that is, it is unavailable to form reducing potential.

From this it can be seen that an efficient storage of energy by the photochemical reduction of DPN^+ is impossible if the DPNH utilizing reactions have activation energy barriers of 14 kcal or greater.

If we, in addition, take into account the fact that oxygen must arise from the precursor formed at the same time as the reductant (energy dissipated in this reaction is not available for storage as reduction potential) and the fact that the DPNH radical probably would require more than 25.5 kcal for its production from DPN^+ photochemically, we see that serious doubts may exist as to the photochemical reduction of pyridine nucleotides in photosynthesis.

A third fact which argues against the photochemical formation of DPNH is the observation that efficient Hill reagents generally possess oxidation-reduction potentials less reducing than the hydrogen zero, or about a quarter of a volt more oxidizing than the $DPN^+/DPNH$ couple.

We have attempted to resolve these difficulties experimentally. If pyridine nucleotides contained in intact plants are reduced at a high rate during the first few seconds of illumination, photochemical reduction seems most likely, despite the above objections. If, on the other hand, they are only slowly reduced or lag behind the rate of CO_2 fixation or ATP formation, for example, then it would appear most likely that the changes are not induced photochemically, but rather that they reflect fluctuations in the size of metabolic pools in equilibrium with the nucleotides.

In cooperation with Dr. V. Lynch at the Carnegie Institution of

Washington, an attempt was made in 1955 to measure changes in the concentration of reduced pyridine nucleotides in Chlorella during illumination. The absorption at 340 mμ was measured with the sensitive difference spectrophotometer used in the experiments described above. We were unable to observe any significant or reproducible changes in transmission at this wave length as a result of the illumination of Chlorella.

Similarly, we attempted at the University of Chicago, to utilize the response of the enzyme, bacterial luciferase, to follow the level of DPNH in Chlorella extracts. However, powerful inhibitors of bacterial extract luminescence exist in the Chlorella extracts, and these experiments were therefore discontinued.

Recently, Mr. A. Gene Ferrari and the author have developed a technique for the assay of DPNH in intact cells, which method utilizes the fact that the reduced form of the coenzyme is fluorescent when it is illuminated with ultraviolet light, whereas the oxidized form is not fluorescent (unpublished; see Fig. 8).

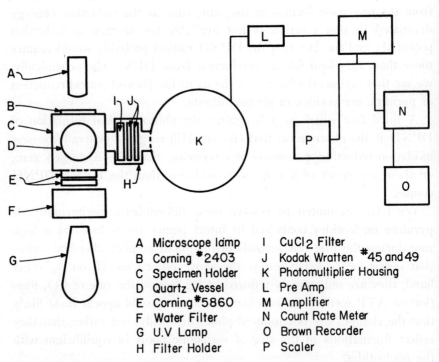

A	Microscope lamp	I	CuCl$_2$ Filter
B	Corning *2403	J	Kodak Wratten *45 and 49
C	Specimen Holder	K	Photomultiplier Housing
D	Quartz Vessel	L	Pre Amp
E	Corning *5860	M	Amplifier
F	Water Filter	N	Count Rate Meter
G	U.V Lamp	O	Brown Recorder
H	Filter Holder	P	Scaler

Fig. 8. Apparatus used to measure changes in reduced pyridine nucleotide concentration in intact plants during photosynthesis.

Contrary to expectations, it was found that illumination in the presence of CO_2 resulted in no increase in fluorescence ascribable to DPNH. On the other hand, following illumination there was a marked decrease in DPNH followed by a slow recovery. In the absence of CO_2, there was, by contrast, a slight increase in DPNH during illumination. When the light was extinguished the fluorescence returned to the base level without overshoot. These results are illustrated in Figs. 9 and 10.

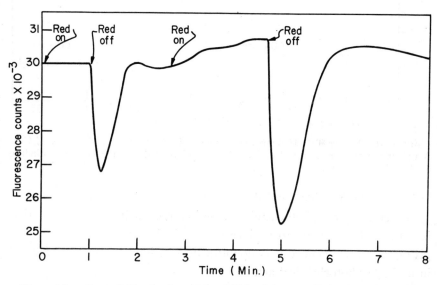

Fig. 9. The effect of illumination with red light (photosynthesis-promoting) on the fluorescence of Chlorella at 480-490 mμ. The fluorescence is proportional to reduced pyridine nucleotide concentration.

Confirmation of the observation that illumination actually results in a net oxidation of pyridine nucleotides has been furnished by the recent experiments of Chance and Sager who showed that illumination decreases the absorption at 340 mμ in the pale mutant of Chlamydomonas discussed earlier (personal communication).

It appears highly improbable that pyridine nucleotides are the natural acceptors of hydrogen in the primary photochemical reaction in photosynthesis. The nature of their participation in the process, if they do participate except very indirectly, is unsettled.

B. Photosynthetic phosphorylation. In the prior section it was pointed out that the primary reductant must give rise eventually to reducing material at approximately the potential of the pyridine nucleotide system, so that this reductant may then itself react reversibly with

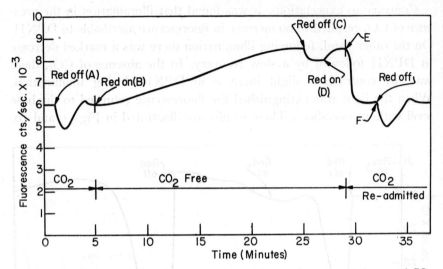

Fig. 10. DPNH or TPNH fluorescence changes in the presence and absence of CO_2. A red light was turned on and extinguished at points *A* through *F* in order to induce or stop photosynthesis.

PGA, ATP, and triose, and thus produce carbohydrate from fixed CO_2.

Either the first photochemical reaction results in a compound that is sufficiently reducing to achieve this or it does not. For the reasons given earlier we believe it is doubtful that such a compound is formed in a single photochemical act.

If we postulate that the photochemical reaction results in a substance whose reduction potential is intermediate between that of water and triose (or DPNH), then it is apparent that photosynthetic systems must contain some device for raising the reduction level of the first reducing substance to that of triose. In principle, only two types of mechanism are available for the generation of higher potentials from lower ones:

(1) The primary reducing agent is made more reducing by absorbing energy from a second quantum, *or*
(2) The reducing potential is increased in one molecule at the expense of the energy in another molecule.

There is at present no evidence suggesting a mechanism embodying the first alternative. The second mechanism is essentially an energy dismutation process and one possible scheme for it is illustrated in Figure 11.

[186]

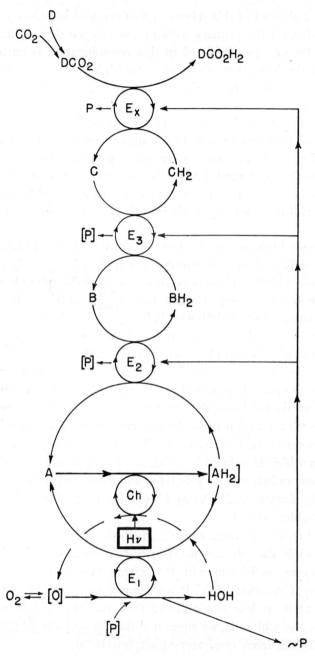

Fig. 11. Ch = photochemical apparatus. Hν = light. AH$_2$, BH$_2$, CH$_2$ = intermediate reductants. [O] = photochemically produced oxidant, precursor to O$_2$. D = "C$_2$ acceptor" molecule. $\sim$ P = ATP. DCO$_2$H$_2$ = primary fixation-reduction product. E$_1$ = terminal oxidase. E$_2$, E$_3$, E$_x$ = intermediate transhydrogenases. From Strehler (1953).

The first element of this scheme is that one unit of primary reductant reacts back with the primary oxidant (or oxygen derived therefrom), and that the energy dissipated in this recombination is partially conserved by the formation of one or more high energy phosphate bonds (or some equivalent non-redox energy storage system in equilibrium with $\sim P$).

Next, one of the high energy phosphate bonds is used to raise the potential of electrons present in the primary reductant to the level of the pyridine nucleotide system. In other words, reducing material at an intermediate potential (AH_2) is used as a source of high energy phosphate bonds ($\sim P$). These bonds in turn are used to increase the reducing ability of another molecule of AH_2 successively to BH_2 and CH_2, etc.

Just how a high energy phosphate bond could be used to increase the reducing ability of another molecule might best be viewed as a formal reversal of oxidative phosphorylation. In oxidative phosphorylation a phosphate bond is formed when reducing potential is lost, i.e. the potential energy in weakly bound hydrogens is exchanged for potential energy in weakly bound phosphates. If this process is reversed the energy in weakly bound phosphates (high energy) is traded for energy in weakly bound hydrogens. This postulated process, the reverse of oxidative phosphorylation, could be called reductive dephosphorylation.

While we do not know the exact mechanism of this process, or even whether such a reaction is involved in photosynthesis, we do know that light does cause the formation of ATP in intact Chlorella. This ATP formation which is induced by light, or photosynthetic phosphorylation as it is now called, was first detected and measured through the use of the Firefly Enzyme ATP Assay (McElroy, 1947; Strehler and Totter, 1952; Strehler, 1953).

Typical results obtained with this method, are shown in Figs. 12 and 13, and while they do not necessarily rule out alternative interpretations, they are consistent with the scheme proposed above. Since the respiratory formation of ATP was shown to be inhibited to a much greater extent at low temperatures than was the light-induced phosphorylation of Chlorella, we suggested that photosynthetic phosphorylation is distinct from respiratory phosphorylation.

Recently Arnon and collaborators (1954) have found that isolated chloroplasts will carry on an active light-induced phosphorylation. These important results, which localize the site of photosynthetic phos-

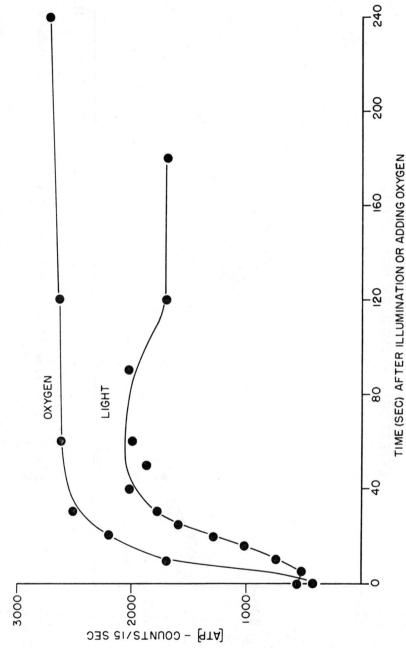

Fig. 12. Effect of light and O₂ admission in the dark on ATP levels in plants kept anaerobic for 2-3 hours. From Strehler (1953).

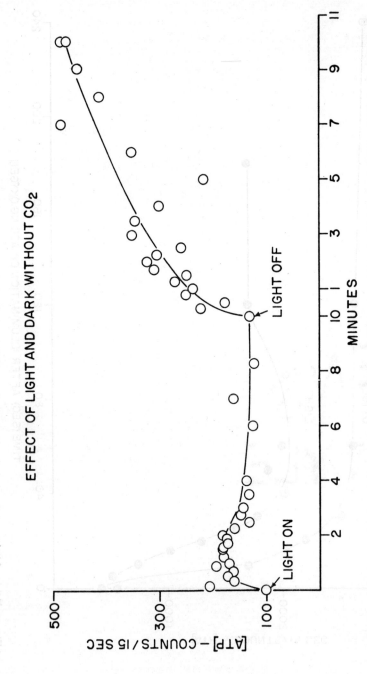

Fig. 13. Effect of light and subsequent dark periods on ATP concentration in the absence of CO₂. (Effects are similar in light in presence or absence of CO₂.) From Strehler (1953).

phorylation in the chloroplast, suggest (as do Frenkel's 1954 studies on phosphorylating particles from photosynthetic bacteria) that photosynthetic phosphorylation takes place without the intervention of free oxygen. Rather, ATP formation can be visualized as arising from the recombination of light-produced oxidizing and reducing substances, perhaps even the primary photochemical products in photosynthesis.

We do not know the molecular species involved in the generation of ATP in either photosynthetic phosphorylation or oxidative phosphorylation. The possible mechanisms which could account for the stepwise transfer of hydrogens or electrons through a series of energy-yielding steps in such a way that part of the energy released is conserved in the form of phosphoric anhydride bonds, can all be broken down into variations on three basic parameters describing the reaction. These variable parameters are the *time relation, spatial relation*, and *energy transfer mechanism* involved.

Time: There are three possible sequences: The anhydride bond energy is conserved (1) before, (2) at the same time, and (3) after the oxidation reduction reaction takes place.

Space: Three possibilities exist: The energy source and energy acceptor may (1) be in direct contact (including the formation of a compound or complex between the redox carrier and the phosphate system), (2) may be in contact through an intermediate material structure, or (3) may be physically isolated from each other.

Energy transfer mechanism: Among the ways in which energy could be transmitted from the redox system to the phosphate system are: (1) formation of unstable bonds by desaturation or rearrangement of covalent bonds between donor and acceptor complex, (2) interaction of electrostatic fields, (3) interaction of magnetic fields, or (4) transfer by electromagnetic radiation.

C. Photosynthetic phosphorylation models. More than 30 possible combinations of variations in these parameters exist. Some of these combinations, and schemes derived from them, are given below. The combination of the three parameters which make up the scheme are indicated in parentheses. Thus in Model 1 the phosphate bond energy is conserved at the same time (2) as the redox energy is liberated, the donor and acceptor are in direct contact (1) and the energy is transferred by mechanism (1). The code for this mechanism is then (2,1,1).

MODEL 1: Based on substrate-linked phosphorylation mechanisms (2,1,1)

$$
\begin{array}{ccc}
\text{(a)} & & \text{(b)} \\
\overset{\diagdown}{\underset{|}{\text{AH}}} & \overset{\diagdown}{\underset{|}{\text{AH}}} & \overset{\diagdown}{\text{A}} \\
\overset{|}{\text{BH}} + \text{P} \rightleftarrows & \overset{|}{\text{BH}} & \overset{\|}{\text{B}} \\
\overset{|}{\text{C}} & \xrightarrow{-2\text{H}} & \overset{|}{\text{C}} \\
\diagup \diagdown & \diagup \diagdown & \diagup \diagdown \\
\text{X} & \text{P} & \sim\!\text{P}
\end{array}
$$

where A, B and C are atoms in the hydrogen carrier and wherein the phosphate bond formed in reaction (a) is tight (i.e. low energy) and that formed as a result of oxidation of the complex is loose (high energy). Variations on this model include the case in which the phosphate is bound tightly in the oxidized state and more loosely (high energy) in the reduced state. Moreover, P could be replaced by some other group which is itself in equilibrium with P.

MODEL II: Based on a Kornberg type reaction (1,1,1)

$$
\text{A}-\text{P} + \text{P}-\text{BH}_2 \; \underset{\text{energy}}{\overset{\text{thermal}}{\underset{\longleftarrow}{\longrightarrow}}} \; \text{A}-\text{P}\sim\text{P}-\text{BH}_2
$$

$$
\text{A}-\text{P}\sim\text{P}-\text{BH}_2 \; \xrightarrow{\;\longleftarrow\;} \; \text{A}-\text{P}\sim\text{P}-\text{B} + 2\text{H}
$$

$$
\text{A}-\text{P}\sim\text{P}-\text{B} + \text{P} \rightleftarrows \text{A}-\text{P}\sim\text{P} + \text{PB}
$$

$$
\text{P}-\text{B} + \text{CH}_2 \rightleftarrows \text{P}-\text{BH}_2 + \text{C}
$$

In this scheme, A—P might be AMP, ADP, etc., and BPH_2 might be reduced nicotinic mononucleotide, for example. Note that the anhydride bond is formed before the electron transfer takes place. The pyrophosphate bond in the dinucleotide (formed by borrowing thermal energy from the environment) is stabilized by the rapid selective oxidation of the newly formed dinucleotide. Subsequently, the stabilized, oxidized dinucleotide is converted into ATP or ADP by a Kornberg type reaction.

MODEL III: Szent-Gyorgyi type (1,3,4)

$$
\text{AH}_2 + \text{C} \rightarrow \text{CH}_2 + \text{A}^* \rightarrow \text{A} + h\nu
$$

$$
h\nu + \text{B}-\text{D}-\text{P} \rightarrow \text{B}-\text{D}-\text{P}^*
$$

$$
\text{B}-\text{D}-\text{P}^* \longrightarrow \text{P}\sim\text{B}-\text{D}.
$$

In this scheme, an energy-yielding oxidation reduction reaction results in luminescence (in the infrared)which in turn excites another molecule containing phosphate to rearrange and form a high energy

[192]

bond from a low energy bond. Essentially, this model consists of a coupling between a bioluminescence and "photosynthesis."

MODEL IV: Conduction band type (1,2,2)

This model was developed in the course of discussions with Dr. Henry Mahler of the University of Indiana.

Consider a protein-lipid matrix in which electrons can move relatively freely from one place to another. On this surface there are potential humps and valleys of increasing depth along some coordinate which are produced by the adsorption of charged positive and negative ions and by the charged ionic character of the protein. The centers of charge are capable of binding oppositely charged groups. These conditions are illustrated in Fig. 14a.

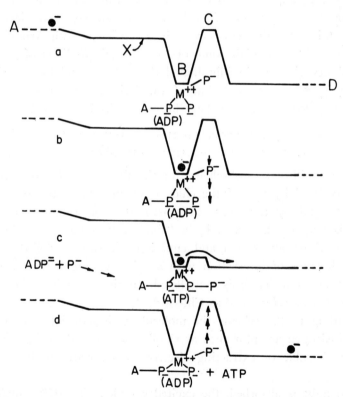

Fig. 14. A scheme for oxidative phosphorylation utilizing conduction bands. See text for details.

Suppose that an electron is released into this conducting system at A from some source. The electron will traverse the "conductor" X until it reaches the potential valley at B, where it will be blocked because of the attraction between the negative charge on the electron and the positive charge which creates the valley, and the repulsion between the negative charges on the hump and the electron (Fig. 14b). The negative charge on the hump (c) is caused by the inherent charge on the structure added to the negative charge on a phosphate bound to the metal ion, M, by chelate bonds.

If there is now a thermally induced dehydration (Fig. 14b, c) between the phosphate and ADP (to yield ATP) the charge distribution will change so as to open the valve and permit the electron to move further along the structure (Fig. 14d). The valve is the charged hump which prevents the movement of the electron beyond c.)

While it is impossible to decide at this time among the above or other possible mechanisms of oxidative phosphorylation, the last model has a certain attractiveness because it is consistent with Arnold's data on photoconduction, with some of Commoner's findings of magnetic susceptibility changes in illuminated chloroplasts and because it furnishes a possible interpretation of the light-induced absorbency increases in Chlorella cells that have been partially inactivated by heat.

IV. DISCUSSION: A PROPOSED MECHANISM LINKING THE PHOTO-CHEMICAL AND BIOCHEMICAL PARTS OF PHOTOSYNTHESIS

Figure 15 illustrates a scheme for the first part of photosynthesis. Light absorbed by any of the chlorophyll molecules in the "unit" is passed around by a process of sensitized fluorescence. The energy finally is trapped, however, by a "sink" consisting of a chlorophyll molecule whose absorption and emission are shifted slightly toward the red because of interaction with another conjugated system (e.g. carotenoid) to which it is complexed. In the following, the excited complex will be referred to as Chl—C*.

According to the scheme the molecule C is associated at one end with a hydrogen acceptor such as riboflavin and at the other end with a system capable of extracting electrons from water in its oxidized state.

When light is absorbed, the excited complex Chl—C* transfers an H to the hydrogen acceptor (X) which then diffuses away to a site of utilization. This leaves the conjugated molecule X in the radical form

$$CH_2 = --- C — CH_2 \cdot$$
$$H$$

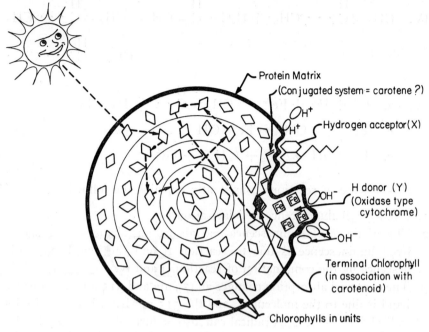

Fig. 15. A hypothetical model relating light absorption and oxidation-reduction reactions at the onset of photosynthesis. See text for details.

This radical ($C^{R.}$) is capable of resonating thus:

$$CH_2 = ---C-CH_2 \cdot \rightleftarrows \cdot CH_2 ---C=CH_2$$
$$H H$$

Since the molecule is a radical, it will tend to extract electrons from potential donors (i.e. it is an oxidant) and since it resonates, it will act as an oxidant at all points on its surface, since the unpaired electron is not localized.

The reactions in which $C^{R.}$ then participates are outlined below:

I. $$CH_2 = \overset{H}{C} \cdots CH_3 + X + h\nu \rightarrow CH_2 = \overset{H}{C} \cdots CH_2 \cdot + XH$$

II. $$CH_2 = \overset{H}{C} \cdots CH_2 \cdot + XH \rightarrow CH_2 = \overset{H}{C} \cdots CH_3{}^* + X$$

III. $$CH_2 = \overset{H}{C} \cdots CH_2 \cdot + P_i + XH \rightarrow$$
$$CH_2 = \overset{H}{C} \cdots CH_3 + X + \sim P$$

IV. $\quad CH_2 = \overset{H}{C} \cdots CH_2 \cdot + H_3C - \overset{H}{C} = CH_2 \rightleftarrows CH_2 = \overset{H}{C} \cdots CH_3$

$$+ CH_2 = \overset{H}{C} \cdots CH_2 \cdot$$

V. a. $\quad C^R + H^+ + Fe^{++} \rightleftarrows CH_2 = \overset{H}{C} \cdots CH_3 + Fe^{+++}$

b. $\quad Fe^{+++} + OH^- \rightarrow Fe^{++}OH$

c. $\quad Fe^{++}OH \rightarrow \frac{1}{4}O_2 + \frac{1}{2}H_2O + Fe^{++}$

This scheme of reactions consistently explains the following:

1. Chlorophyll fluorescence depolarization.
2. The lifetimes of luminescence and absorption changes. The short-lived luminescence is due to reaction II and the long-lived luminescence by the combination of reaction II and IV.
3. The observed absorption changes at 475 mμ and 515 mμ. The former band is due to the molecule C and the 515 mμ band is caused by its radical form, C^R. This radical can also be formed oxidatively by the reverse of reaction v. This accounts for the spectral changes observed by Chance and Strehler (p. 178).
4. The parallel decay of conduction and luminescence of dried chloroplasts occurs through reaction IV as follows:

$$XH \rightarrow X + H^+ + e^-$$

$$\cdot CH_2 - - - \underset{H}{C} = CH_2 + e + H^+ \rightarrow CH_3 - - - \underset{H}{C} = CH_2$$
$$\text{(chain breaking)}$$

V. CONCLUSION

Photosynthesis will probably prove to have little or no resemblance to the picture of its unknown parts here presented. But, it does now appear likely that the relatively simple and direct schemes, such as the photochemical reduction of fixed CO_2, which have attracted some physicists working in the field and the direct photochemical reduction of pyridine nucleotides by "splitting water," which has been proposed from the biochemists' side, are inadequate to account for the experimental results.

The present evidence suggests that DPNH participates only very indirectly in photosynthesis, if at all, and that the oxidation of this substance which is observed following a period of photosynthesis is caused by an increased rate of oxidative phosphorylation following illumination because of the accumulation of acceptors for phosphate in the light (compare Fig. 9 and Fig. 13).

It may well be that the reduction of PGA to triose in photosynthesis does not involve the pyridine nucleotides at all, but that the electrons or hydrogens arrive at the enzymatic site of PGA reaction by being conducted through the structure of the chloroplast rather than by being first attached to a diffusable carrier. It is likewise possible that photosynthetic phosphorylation takes place without the participation of diffusable molecules other than phosphate and phosphate acceptor and that the phosphorylative reactions involve electron migrations in semisolid structures.

The complexities of photosynthetic systems undoubtedly have their parallels in problems of growth and development. The mechanisms of energy transduction and transfer which have survival value in photosynthetic systems will probably also be found useful to non-photosynthetic organisms. It is because the direct chemical attack on problems in photosynthesis has been unrewarding in the areas here discussed, that optical techniques have been developed and applied. If and when these and similar physical and optical methods are applied to the study of energy transductions in growth and embryogenesis, the results may be even more confusing, though hardly less interesting subjectively or objectively.

BIBLIOGRAPHY

Arnold, W. A., and J. R. Oppenheimer. 1950. Internal conversion in the photosynthetic mechanism of blue-green algae. *J. Gen. Physiol. 33*, 423-435.

Arnold, W. A., and J. B. Davidson. 1954. The identity of the fluorescent and delayed light emission spectra in *Chlorella. J. Gen. Physiol. 37*, 677-684.

Arnon, D. I. 1951. Extracellular photosynthetic reactions. *Nature 167*, 1008-1010.

Arnon, D. I., F. R. Whatley, and M. B. Allen. 1954. Photosynthesis by isolated chloroplasts. II. Photosynthetic phosphorylation, the conversion of light into phosphate bond energy. *J. Am. Chem. Soc. 76*, 6324-6329.

Arthur, W. E., and B. L. Strehler. Studies on the primary process in photosynthesis. I. Photosynthetic luminescence: multiple reactants. *Arch. Biochem. Biophys.* (in press).

Bassham, J. A., A. A. Benson, L. D. Kay, A. Z. Harris, A. T. Wilson, and M. Calvin. 1954. The path of carbon in photosynthesis. XXI. The cyclic regeneration of carbon dioxide acceptor. *J. Am. Chem. Soc. 76*, 1760-1770.

Bruegger, J. 1954. Thesis, University of Chicago.

Coleman, J. W., A. S. Holt, and E. Rabinowitch. 1956. Reversible bleaching of chlorophyll in vivo. *Science 123*, 462-463.

Duysens, L. N. M. 1954. Reversible changes in the absorption spectrum of *Chlorella* upon irradiation. *Science 120*, 353-354.

Ferrari, A. G., W. E. Arthur, and B. L. Strehler. Studies on chlorophyll chemiluminescence. In Gatlinburg Photosynthesis Conference Proc. (in press).

Franck, J. 1953. Participation of respiratory intermediates in the process of photosynthesis as an explanation of abnormally high quantum yields. *Arch. Biochem. Biophys. 45*, 190-229.

Frenkel, A. 1954. Light induced phosphorylation by cell-free preparations of photosynthetic bacteria. *J. Am. Chem. Soc. 76*, 5568-5569.

Krasnovsky, A. A. 1948. Obratimoe fotokhimicheskoe vosstanovlenie khlorofilla askorbinovoĭ kislotoĭ. *Doklady Akad. Nauk SSSR 60*, 421-424.

Linschitz, A., and J. Rennert. 1952. Reversible photo-bleaching of chlorophyll in rigid solvents. *Nature 169*, 193-194.

Livingston, R., and V. A. Ryan. 1953. The phototropy of chlorophyll in fluid solutions. *J. Am. Chem. Soc. 75*, 2176-2181.

Lundegårdh, H. 1954. On the oxidation of Cytochrome f by light. *Physiologia Plantarum 7*, 375-382.

McElroy, W. D. 1947. The energy source for bioluminescence in an isolated system. *Proc. Nat. Acad. Sci. U.S. 33*, 342-345.

Strehler, B. L. 1951. The luminescence of isolated chloroplasts. *Arch. Biochem. Biophys. 34*, 239-248.

Strehler, B. L. 1952. Photosynthesis—energetics and phosphate metabolism. In *Phosphorus Symposium*, Vol. 2, p. 491, ed. by W. D. McElroy and H. B. Glass. The Johns Hopkins Press, Baltimore.

Strehler, B. L. 1953. Firefly luminescence in the study of energy transfer mechanisms. II. Adenosine triphosphate and photosynthesis. *Arch. Biochem. Biophys. 43*, 67-79.

Strehler, B. L., and W. Arnold. 1951. Light production by green plants. *J. Gen. Physiol. 34*, 809-820.

Strehler, B. L., and J. R. Totter, 1952. Firefly luminescence in the study of energy transfer mechanisms. I. Substrate and enzyme determination. *Arch. Biochem. Biophys. 40*, 28-41.

Strehler, B. L., and V. H. Lynch. 1956. Photosynthetic luminescence and photoinduced absorption spectrum changes in *Chlorella. Science 123*, 462-463.

Strehler, B. L., and V. H. Lynch. Studies on the primary process in photosynthesis. II. Some relationships between light induced absorption spectrum changes and chemiluminescence during photosynthesis. *Arch. Biochem. Biophys.* (in press).

Tolmach, L. J. 1951. The influence of triphosphopyridine nucleotide and other physiological substances upon oxygen evolution from illuminated chloroplasts. *Arch. Biochem. Biophys. 33*, 120-142.

Vishniac, W., and S. Ochoa. 1952. Fixation of carbon dioxide coupled to photochemical reduction of pyridine nucleotides by chloroplast preparations. *J. Biol. Chem. 195*, 75-93.

Witt, H. 1955. Kurzzeitige Absorptionsänderungen beim Primärprozess der Photosynthese. *Naturwiss. 42*, 72-73.

X. PHOTOSYNTHESIS AND LIFE ON
OTHER PLANETS: A SUMMARY

BY HARLOW SHAPLEY[1]

MANY laboratories are at present studying the problems of the origin and first development of life. The sciences concerned are biochemistry, geophysics, atomic physics, cytology, paleontology, cosmography, in fact, nearly all disciplines. Even the astronomer contributes from his abundance of years and his steady stores of stellar energy.

First, let us note the places where there is no photosynthesis. We can begin with the moon, which is waterless and has no vegetation whatsoever. Its atmosphere is almost totally lacking; all of the lightweight atoms escaped long ago because of the moon's small mass and consequential low surface gravity. (It has a very small amount of argon, much less than that in the earth's atmosphere, where it is about 1%.)

There is no life on comets, for they are simply "gravel banks in orbital motion," having no water and, of course, no photosynthesis.

There can be no life on the sun. Some speculators in the past have considered the possibility of solar biology; and Herschel had a hypothesis that there is life under the surface of the sun, especially where the sunspots indicated cooler areas. Present-day astrophysics has dispelled any such belief. The sun and stars, with temperatures in the thousands of degrees, are hostile to any but the most simple molecules. There is little chance of life in the centers of globular clusters (Plate I). The stars are much too close together, and gravitational perturbations would wreck planetary orbits.

We do not know for certain the origin of the planets, on the surfaces of which there is a possibility of photosynthesis and life. At least fifteen different theories concerning their origin can be listed (Table I). No one theory has been proven entirely correct, and some are undoubtedly entirely incorrect. It is reasonable to assume that existing systems started in various ways, their origin involving aboriginal explosions, collisions, near misses, rotational fission, double-star troubles, and shrinking gaseous nebulae.

Five hundred years ago, with the unreinforced human eye as our only tool, we saw only about 3000 stars. We naturally thought we were in the center of the stellar system. With modern telescopes we can now

---

[1] Department of Astronomy, Harvard University.

TABLE I. Fifteen hypotheses of planetary origin

---

1. The Mosaic cosmogony, and similar early doctrines of religion.
2. Nebular hypothesis, the famous long-enduring Kantian-Laplacian theory.
3. Partial disruption of the sun by a comet, with the production of planets.
4. Solar eruptions providing planet-building "planetesimals."
5. Capture of the planets by the sun from space or from other stars.
6. Tidal disruption of the sun by a passing star, providing gaseous filaments that condense into planets (variant of 4).
7. Glancing collision of stars (variant of 6).
8. Break-up of one component of a binary star by a third passing star.
9. Explosive fission of the hypothetical proto-sun.
10. Disruption of an unstable pulsing variable star (cepheid).
11. Revival of the nebular hypothesis, bolstered by modern theories of dust and gas accretions.
12. Electromagnetically produced condensations in a contracting nebula (variant of 2).
13. Nova explosion in a binary system providing circulating planetoidal fragments.
14. Revival of the hypothesis of cold planetesimals operating in a nebulous medium (combination of variants of 4, 11, and 12).
15. Primeval explosive chaos and the Survival of the Conforming.

---

see millions, and photograph billions, each one a potential source of life in that it provides energy which can activate the life-producing biochemical processes. Certain conditions must be present, however, if planetary life is to develop. There must be: (1) water in a liquid form, (2) a suitable rotation period, (3) low orbital eccentricity, (4) favorable chemistry of air, water, and rock, (5) a steady star as an energy source, and, of decisive importance, (6) life must get started. But the universe contains abundant cosmic food for the development of life, since high-energy radiation, oxygen, nitrogen, hydrogen, carbon, and other essential elements like phosphorus, calcium, potassium, magnesium, and iron are found everywhere throughout the universe.

Our Milky Way system is a large galaxy, discoidal in shape. The planetary system in which we are located is definitely "peripheral," 25,000 light years from the center, near the edge of our one galaxy among the billions. We now have, from our telescopes and cameras, a pretty good idea of the size and proportions of our local galaxy. We measure the motions of others and find that all are receding from each other. From this fact we have the concept of the expanding universe.

Four to six thousand million years ago the galaxies were crowded together, and our planet was formed during that time of high density. Among catalogued stars there are about 40,000 like our sun, and the probability is high that a majority of them had similar experiences in past ages—collisions, explosions, contractions, etc. There are, therefore, probably not less than 100,000 million planetary systems in the billions of galaxies.

In the Metagalaxy, which is the name given to the universe of galaxies within our range, there are, conservatively, 10^{20} stars, each a source of energy which could support life on any suitable planet that may accompany it. Given the proper conditions (chemical and physical), life appears inevitable. While we might not recognize such life if we saw it, the more I read, hear, and think about it, the more I am convinced that life anywhere should not be too different from that on earth.

In view of the conditions necessary to produce life, it becomes apparent that only Mars, Earth, and Venus of the planets in our solar system are within the proper temperature range. Mercury is much too hot, being closest to the sun. The atmosphere of Venus consists largely of carbon dioxide, with no measurable amount of oxygen or water. It is therefore not a good prospect for the support of life. However, we cannot be certain, because the surface of Venus is concealed by impenetrable dusty and icy atmospheres. It has been studied thermometrically recently; the surface appears to be hot and its atmosphere cold.

The first atmosphere of the earth contained little free oxygen; it was mostly methane, ammonia, nitrogen, hydrogen, and oxides of hydrogen. Sources of energy that helped in the early natural synthesis were the body heat of the earth (manifested in geysers and volcanos), ultraviolet radiation from the sun, ultra-ultraviolet radiation (gamma rays) from radioactive decomposition of certain elements, and lightning discharges through the primeval gases, all of which could activate the essential biochemical reactions.

Various photographic studies of Jupiter indicate that its atmosphere now consists chiefly of methane and ammonia. Radio telescopes in Washington a few years ago picked up radio signals from the planet, and people began to wonder about the possibility of Jovian "hams." These radio waves are being studied with care at the Boulder (Colorado) observatory of the National Bureau of Standards. They appear to be caused by electrical storms whose frequencies, locations, and depths can be measured. (Studies of this sort come in the category of Radio-astronomy, a relatively new science. There is now in active operation at

Harvard a 24-foot radio telescope; one of 60-foot aperture has just been dedicated. The primitive lightnings in these Jovian storms must be very powerful, and the "Thunderbolts of Jove" may be a good name for the radio signals. The winds and turbulence on that planet are tremendous. On the whole, it is most unfavorable as a location for living organisms—too cold, too windy, and smelly of noxious gases (Fig. 1).

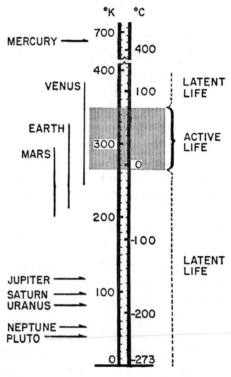

Fig. 1. The planetary thermometer. (Reproduced by permission from *The Green and Red Planet*, by Hubertus Strughold, 1953, copyright by The University of New Mexico Press, Albuquerque.)

In September of 1956, precisely on schedule, Mars came close to the earth and was subject to careful observation. It is hoped that when the observations are reduced new things will be learned about it.[2] The latest theory on the nature of the "canals" on Mars is that they are natural surface faults, not connected but from a distance appearing to be continuous. American observers feel satisfied that the "canals" exist; European observers, on the other hand, believe that they are certainly

[2] Later: Radio signals were detected, and thousands of photographs were made.

PLATE I. In a globular cluster there is too much crowding for the slow evolution of biological chemistry.

discontinuous and, moreover, they do not believe that the "canals" have been artificially constructed. Nobody does now, for that matter.

The surface of Mars is difficult to study because the planet is small and frequently clouded with dust. Photographs made in the light of different colors give some idea of the physical nature of the dust. The density of the Martian atmosphere is about equal to that twelve miles above the earth's surface; but eighteen miles above the Martian surface the atmosphere is denser than that of the earth at the same height. The planet is thus well protected from the impact of meteors. The moon has no such security.

In the opinion of James Franck the composition of the atmosphere of Mars (carbon dioxide, nitrogen, argon, and other inert gases), and the essential absence of water in liquid form, would indicate that photosynthesis on Mars is not as it is on the earth; but this does not preclude the possibility that some forms of life do exist there. The polar caps, probably of hoar frost, indicate that a trace of moisture is present. But if there is life, it must be somewhat different from ours and probably of low estate.

INDEX FOR THE FIFTEENTH GROWTH SYMPOSIUM

CUMULATIVE AUTHOR INDEX

LIST OF GROWTH SYMPOSIA

Roman numerals in the index refer to symposium number.

	YEAR	PLACE OF MEETING	PUBLICATION
I	1939	North Truro, Mass.	*Growth*, Vol. 3, Supplement
II	1940	Salisbury Cove, Me.	*Growth*, Vol. 4, Supplement
III	1941	Hanover, N.H.	*Growth*, Vol. 5, Supplement
IV	1942	North Truro, Mass.	*Growth*, Vol. 6, Supplement
V	1945	North Truro, Mass.	*
VI	1946	Kingston, R.I.	*Growth*, Vol. 10, Supplement
VII	1947	Storrs, Conn.	*Growth*, Vol. 11, pp. 194-358
VIII	1948	Burlington, Vt.	*Growth*, Vol. 12, Supplement
IX	1949	New London, Conn.	*Growth*, Vol. 13, Supplement
X	1951	Northampton, Mass.	*Growth*, Vol. 15, Supplement
XI	1952	Williamstown, Mass.	*Dynamics of Growth Processes*, 1954**
XII	1953	Durham, N.H.	*Biological Specificity and Growth*, 1955**
XIII	1954	Hanover, N.H.	*Aspects of Synthesis and Order in Growth*, 1955**
XIV	1955	Amherst, Mass.	*Cellular Mechanisms in Differentiation and Growth*, 1956**
XV	1956	Providence, R.I.	*Rhythmic and Synthetic Processes in Growth*, 1957**

Adolph, E. F. Discussion remarks to the paper of Herbert Freundlich. II, 52-54

Allen, Edgar. The rates of growth in genital tissues and the hormonal regulation involved. IV, 73-82

Avery, George S., Jr. Chemical factors of plant growth. II, 55-72

Beadle, G. W. *See* Tatum

Berrill, N.J. Spatial and temporal growth patterns in colonial organisms. III, 89-111

Bessey, Otto A. Tissue responses to vitamin deficiencies. IV, 95-104

Billingham, R. E. Acquired tolerance of foreign cells. XIV, 221-232

Black, L. M. Virus tumors in plants. VI, 79-84

Blakeslee, Albert F. Growth patterns in plants. III, 77-88

Blum, Harold F. On the origin of self-replicating systems. XV, 155-170

Bodenstein, Dietrich. Humoral agents in insect morphogenesis. XIII, 257-268

Boell, E. J. Biochemical and physiological analysis of organizer action. IV, 37-53

Bonner, James, and Sam G. Wildman. Contributions to the study of auxin physiology. VI, 51-68

Borthwick, H. A. *See* Hendricks

Brachet, Jean. Biochemical and physiological interrelations between nucleus and cytoplasm during early development. VII, 309-324

Braun, Armin C. Recent advances in the physiology of tumor formation in the crown-gall disease of plants. VII, 325-337

Briggs, Robert, and Thomas J. King. Specificity of nuclear function in embryonic development. XII, 207-228

Brody, Samuel. Agricultural problems in growth and aging. VI, 69-78

Brown, R. and E. Robinson, Cellular differentiation and the development of enzyme proteins in plants. XII, 93-118

Bruce, Victor G. *See* Pittendrigh

Bünning, E. Endogenous diurnal cycles of activity in plants. XV, 111-126

* This symposium was not published in any form; for the record, the authors and titles of individual papers are included in the present index.

** Symposia XI-XV published by the Princeton University Press.

Holtfreter, Johannes. Some aspects of embryonic induction. x, 117-152

Horowitz, N. H. Genetic and non-genetic factors in the production of enzymes by Neurospora. x, 47-62

Irwin, M. R. Genes, antigens and antibodies. v

Irwin, M. R. Immunogenetics. xii, 55-71

King, Thomas J. *See* Briggs

Klein, Richard M. Growth and differentiation of plant tissue cultures. xv, 31-58

Kozloff, Lloyd M. Virus reproduction and the replication of protoplasmic units. xi, 3-20

Landauer, Walter. Hereditary abnormalities and their chemically-induced phenocopies. viii, 171-200

Lederberg, Joshua, and Esther M. Lederberg. Infection and heredity. xiv, 101-124

Lewis, Warren H. Some contributions of tissue culture to development and growth. i, 1-14

Li, Choh Hao. Growth and anterior pituitary. viii, 47-60

Loofbourow, John R. Effects of ultraviolet radiation on cells. viii, 75-149

Lwoff, A. Kinetosomes and the development of ciliates. ix, 61-91

Maculla, Esther. Antigenic analysis of embryonic, adult and tumor tissues. ix, 33-60

Manton, I. Plant cilia and associated organelles. xiv, 61-71

Mazia, Daniel. Desoxyribonucleic acid and desoxyribonuclease in development. ix, 5-31

McLean, Franklin C. Calcification and ossification. iv, 83-94

Monod, Jacques. The phenomenon of enzymatic adaptation and its bearings on problems of genetics and cellular differentiation. vii, 223-289

Needham, Joseph. Biochemical aspects of organizer phenomena. i, 45-52

Niu, M. C. New approaches to the problem of embryonic induction. xiv, 155-171

Northrop, F. S. C. The method and theories of physical science in their bearing upon biological organization. ii, 127-154

Novick, Aaron, and Leo Szilard. Experiments with the chemostat on the rates of amino acid synthesis in bacteria. xi, 21-32

Pauling, Linus. The duplication of molecules. xiii, 3-13

Pittendrigh, Colin S., and Victor G. Bruce. An oscillator model for biological clocks. xv, 75-109

Pollister, Arthur W. Cytochemical aspects of protein synthesis. xi, 33-67

Porter, Keith R. Cell and tissue differentiation in relation to growth (animals). xi, 95-110

Prescott, David M. Relations between cell growth and cell division. xv, 59-74

Puck, Theodore T. The mammalian cell as microorganism. xv, 3-17

Raper, John R. Some problems of specificity in the sexuality of plants. xii, 119-140

Raper, Kenneth B. Developmental patterns in simple slime molds. iii, 41-76

Rasch, Ellen. *See* Swift

Rebhun, Lionel. *See* Swift

Reichardt, W. *See* Delbrück

Reimann, S. P. Discussion remarks to the paper of H. S. N. Greene. ii, 124-125

Riker, A. R. The relation of some chemical and physico-chemical factors to the initiation of pathological plant growth. iv, 105-117

Robbins, William J. Some factors limiting growth. ix, 177-186

Robinson, E. *See* Brown

Rothen, Alexandre. Long range forces in immunologic and enzymatic reactions. vii, 195-203

Russell, Elizabeth S. Review of the pleiotropic effects of W-series genes on growth and differentiation. xiii, 113-126

Sax, Karl. Incompatibility in selfing and crossing plant species. v

Schechtman, A. M. Ontogeny of the blood and related antigens and their significance for the theory of differentiation. xii, 3-31

Schmitt, Francis O. Some protein patterns in cells. iii, 1-20

Schmitt, Francis O. Seeing the invisible

approach to some problems of induction and differentiation. XII, 33-53

Woodard, John. *See* Swift

Woodger, J. H. Remarks on method and technique in theoretical biology. I, 87-99

Woodger, J. H. Notes on the first sym-posium on development and growth. I, 101-111

Woolley, D. W. The study of growth and metabolism by means of specific antagonists. VIII, 3-16

Wrinch, Dorothy. Discussion remarks to the paper of O. L. Sponsler. II, 26